AF505951

Designing Groundwater Models with WINDOWS™

William C. Walton

LEWIS PUBLISHERS
Boca Raton   Ann Arbor   London   Tokyo

Library of Congress Cataloging-in-Publication Data

Walton, William Clarence
    Designing groundwater models with *Windows* software / by William C. Walton
        p.   cm.
    Includes bibliographical references and index.
    ISBN 1-56670-110-4
        1. Groundwater—Computer simulation.   2. DesignMod.   I. Title.
GB1001.72.E45W33   1995
551.49'01'13543—dc20   94-18439
                                                    CIP

Btrieve is a registered trademark of Novell, Inc.

CompuServe is a registered trademark of CompuServe, Inc.

dBASE, dBASE II, dBASE III, dBASE IV, and dBASE III Plus are registered trademarks of Ashton-Tate Corporation.

Data Widgets, VBAssist, and 3-D Widgets are registered trademarks of Sheridan Software Systems, Inc.

GFA-Basic for Windows Pro Version is a registered trademark of GFA Software Technologies.

High Edit, MUSCLE, and 3-D Gizmos are registered trademarks of MicroHelp, Inc.

Lotus, Freelance, Graphics, Lotus 1-2-3, and DIF are registered trademarks of Lotus Development Corporation.

Harvard Graphics is a registered trademark of Software Publishing Corporation.

Hewlett-Packard, HP, LaserJet, and PCL are registered trademarks of Hewlett-Packard Company.

IBM, DisplayWrite, and OS/2 Extended Edition are registered trademarks of International Business Machines, Incorporated.

Macintosh is a registered trademark of Apple Computer, Inc.

MathCad is a registered trademark of MathSoft.

Micrografx, Micrografx Designer, and Micrografx Draw Plus are registered trademarks of Micrografx Corporation.

Microsoft, Access for Windows, Access Distribution Kit, Draw, Excel for the Macintosh, Excel for OS/2, Excel for Windows, Fortran, Fortran PowerStation, Fortran PowerStation 32, Fox Pro, Graph, MS, MS-DOS, Multiplan, SQL Server, Test for Windows, Visual Basic Application Edition, Visual C++, Visual C++ Professional, Windows, Windows NT, Word for DOS, Word for the Macintosh, Word for Windows, Works for DOS, and Works for Windows are registered trademarks of Microsoft Corporation.

ModelCad 386 is a registered trademark of Geraghty & Miller, Inc.

Oracle is a trademark of Oracle Corporation.

Paintbrush is a trademark of ZSoft Corporation.

Paradox is a registered trademark of Ansa Software, a Borland International, Inc., Company.

PinPoint VB is a registered trademark of Avanti Software, Inc.

PostScript is a registered trademark of Adobe Systems, Inc.

Q+E is a trademark of Pioneer Software Systems Corporation.

QuickPac Scientific for Windows is a registered trademark of Crescent Software, Inc.

Site GIS is a registered trademark of Geo Trans, Inc.

Spread/VBX is a registered trademark of FarPoint Technologies.

Standford Graphics is a registered trademark of 3-D Visions Corporation.

TrackDeck is a registered trademark of Dashboard Software.

Turbo Pascal for Windows, Pascal with Objects, and Visual Solutions are trademarks of Borland International, Inc.

VBCompress is a registered trademark of Tailored PCs.

VB/Magic Controls and Visual/db are registered trademarks of AJS Publishing, Inc.

Visual Tools is a registered trademark of Evergreen Technologies.

WATCOM Fortran is a registered trademark of WATCOM Systems.

WordPerfect and DrawPerfect are registered trademarks of WordPerfect Corporation.

WordStar is a registered trademark of MicroPro International.

3D Graphics Tools is a registered trademark of Microsystem Options.

PREFACE

This book is about designing a groundwater model (conceptualization) and is not about implementing a groundwater model (data preparation with a preprocessor, code execution, interpretation of results). Designing groundwater models is excessively time-consuming and tedious if the modeler has to mechanically draw and redraw many geologic, hydrogeologic, and geochemical cross sections and plan views with overlays. The use of commercial computer-aided design (CAD) systems, geographical information systems (GIS), databases, spreadsheets, graphical applications, mapping applications, mapping interfaces, GIS interfaces, and/or model interfaces can largely solve this problem. The software and digitizers for these systems, however, are expensive (starting at $1000), and most software is not groundwater model design oriented. In addition, users must learn each software package and work with several import/export file formats.

The inexpensive and integrated CAD software *DesignMod* included with this book is customized for designing groundwater models. *DesignMod* does not require a digitizer but instead uses the mouse and the rich graphical features of the *Windows* environment. It enables the user to easily and quickly draw and redraw cross sections and plan views with site base map information, contour, and zonation pattern overlays. *DesignMod* also can be used to manipulate and view model database array files. *DesignMod* supplements *ModelCad* and is less sophisticated than *Site GIS*. *DesignMod* is not a preprocessor for any specific model, although it can be used to create some model database files. It is hoped that *DesignMod* will focus more attention to the design aspects of groundwater modeling and encourage students and professionals to devote additional time to the study of geologic, hydrogeologic, and geochemical data.

Discretization, aggregation, averaging, and approximation guidelines for designing groundwater models with *DesignMod* are provided as a handy reference. *DesignMod* can communicate with commercial wordprocessor, spreadsheet, database management system, and graphics *Windows* software through the *Windows* Clipboard, ASCII text files, and bitmap image files. Features of Microsoft *Word, Excel,* and *Access* and 3-D Vision's *Standford Graphics* that are related to groundwater model design and supplementing *DesignMod* features are described to promote further use of commercial *Windows* software in the groundwater industry.

Graphically oriented *Windows* applications such as *DesignMod* are ideally suited for non-speed critical data entry and display. An example of *DesignMod's* Microsoft *Visual Basic* source code and information about *Visual Basic* programming and using *Windows* functions and Third-Party dynamic link library Add-ons are included in this book to foster greater use of *Windows* programming languages to create groundwater model pre- and post processors and utility programs. *DesignMod* executable and exercise (but not source code) files are provided on a disk distributed with this book.

The User's Manual for *DesignMod* was written with hydrogeologists, geologists, engineers, environmental scientists, soil scientists, governmental agencies, industrial environmental departments, waste management companies, and academia in mind. Many exercises with step-by-step program operation documentation are provided for beginning and intermediate modelers.

THE AUTHOR

William C. Walton received his B.S. degree in Civil Engineering from Lawrence Technological University, Southfield, Michigan, in 1948 and attended Indiana University, University of Wisconsin, Ohio State University, and Boise State University. Bill served for 10 years as Director of the Water Resources Research Center and Professor of Geology and Geophysics at the University of Minnesota. He taught groundwater courses at the University of Illinois on a part-time basis for 2 years.

Bill's 47 years of experience in the water resources field include 2 years as a water well contractor in Detroit, Michigan; 1 year with the U.S. Bureau of Reclamation in Cody, Wyoming; 8 years with the U.S. Geological Survey in Wisconsin, Ohio, and Idaho; 6 years with the Illinois State Water Survey at Urbana, Illinois; 6 years with consulting firms, including Shaefer and Walton, Columbus, Ohio, and Camp Dresser and McKee, Inc., and Geraghty and Miller in Champaign, Illinois; and 3 years as a self-employed consultant of water resources in Mahomet, Illinois. He was executive director of the Upper Mississippi River Basin Commission in Minneapolis, Minnesota, for 5 years and has participated in water resources projects throughout the United States and Canada, in addition to Haiti, El Salvador, Libya, and Saudi Arabia. Bill is presently a self-employed writer.

The positions he has held include Water Resources Planning Director, Minnesota State Planning Agency; Minnesota's member of the Great Lakes Basin Commission, Souris-Red-Rainy River Basins Commission, Upper Mississippi River Basin Coordinating Committee, and Missouri Basin Inter-Agency Committee; advisor to the Land and Water Resources Committee of the Minnesota House of Representatives; member of the Technical Advisory Committee to the Minnesota-Wisconsin Boundary Area Commission; member of the Citizen Advisory Committee, Minnesota Environmental Quality Council; vice president of the National Water Well Association and founding editor of the *Journal Ground Water*; chairman of the Ground Water Committee of the Hydraulics Division; member of the American Society of Civil Engineers; member of the U.S. Geological Survey Advisory Committee on Water Data for Public Use; Consultant to the Office of Science and Technology, Washington, D.C.; member of the steering committee of the International Ground Water Modeling Center; member of the Committee for the International Hydrological Decade; member of the steering committee of the International Field Year on the Great Lakes; and advisor to the United States Delegation to the Coordinating Council of the International Hydrological Decade of UNESCO.

Bill served as a visiting scientist for the American Geophysical Union and the American Geological Institute and has lectured at many universities throughout the United States. He lectured at short courses sponsored for several years by the International Ground Water Modeling Center and has presented papers at several professional society meetings in Europe. Bill is author of over 75 technical papers and eight books: *Groundwater Resource Evaluation* (McGraw-Hill, New York); *The World of Water* (Weidenfeld and Nicolson, London); *Practical Aspects of Groundwater Modeling* (National Ground Water Association); and *Groundwater Pumping Tests, Analytical Groundwater Modeling, Numerical Groundwater Modeling, Principles of Groundwater Engineering,* and *Groundwater Modeling Utilities* (Lewis Publishers, Chelsea, Michigan).

CONTENTS

DEDICATION

This book is dedicated with love to my wife Ellen. For 46 years she has encouraged and supported my endeavors. Besides that, she feeds me well as long as I clean and maintain the house cheerfully. I also have agreed to fold the clothes and help process produce from her garden.

1 INTRODUCTION

Microsoft *Windows (Windows)* has become the graphical user interface (GUI) of choice for many business, scientific, and engineering microcomputer users. *Windows* applications support multitasking, have a consistent user interface, and support superb device-independent graphics. They are more fun to run than DOS applications. Hundreds of word processor, spreadsheet, database, desktop publishing, computer aided design (CAD) graphics, mapping, and utility *Windows* titles are listed in mail-order discount software catalogs. In contrast, few *Windows* titles were listed in groundwater industry catalogs in the first half of 1994.

Number-crunching, memory-hungry, and speed-critical groundwater model programs are developed with compilers that produce 32-bit *MS-DOS* extended applications (see Microsoft Fortran *PowerStation,* released in 1993) or 32-bit *Windows* NT applications (see Microsoft Fortran *PowerStation 32,* released in 1993) instead of *Windows* program development software such as *Visual Basic* that produces slower 16-bit extended applications. Microsoft will release Fortran 90 in the second half of 1994 for *Windows* NT and *Chicago (Windows* 4.0) operating systems. Groundwater model pre- and postprocessors and utility programs are non-speed critical and could be developed with 16-bit *Windows* program development software (see Ribar, 1993, pp. 161–180, *"Using Visual Basic as a Front-End Generator"*). Before 1991, developing *Windows* applications was much harder than developing *DOS* applications. A *Windows* application programmer could expect to learn over 600 functions at a low level and write applications in either assembly or C language. This discouraged the development of groundwater model *Windows* software.

Microsoft introduced *Visual Basic* in 1991, *Visual C++* in 1993, and *Visual C++ Professional* in 1994, which greatly simplified the *Windows* application development process and further encouraged the widespread development and use of *Windows* software. Groundwater model pre- and postprocessor and utility *Windows* programs can now be developed graphically in a fraction of the time needed prior to 1991. *Visual Basic* is general purpose and fast (as fast or nearly as fast as either *C++* or *Turbo Pascal for Windows* in many tests). Its programming language (similar but not identical to

QuickBASIC) has easy-to-use graphics statements, powerful built-in functions for mathematics and string manipulations, and sophisticated file-handling capabilities. *Visual Basic* makes it easy to build large programs using modern modular techniques and lets you interactively place menus, text boxes, command buttons, option buttons, check boxes, list boxes, scroll bars, and file and directory boxes on windows. The screen can contain multiple windows which are able to communicate with one another and with other *Windows* applications running at the same time.

Some professionals utilize specialized software packages to design groundwater models. These packages and associated digitizers usually cost in excess of $1000 and are therefore inaccessible to many professionals as well as students. Information concerning several popular specialized groundwater-oriented software packages is provided below.

ModelCAD 386, distributed by Geraghty & Miller, Inc., 10700 Parkridge Blvd., Suite 600, Reston, VA 22091, ($750), provides a fast, flexible, and interactive method of designing finite-difference groundwater flow and transport models. It interfaces with commercial PC CAD systems and Golden Software, Inc. (P.O. Box 281, Golden, CO 80402) graphics software. *ModelCAD 386* creates data files for the following popular groundwater models: MODFLOW, MODPATH, MOC, and MT3D.

GeoTrans, Inc. (46050 Manekin Plaza, Suite 100, Sterling, VA 20166) distributes ***Site GIS,*** a *Windows* software package for analyzing and presenting environmental data for soil and groundwater remedial investigations. Groundwater, soil, and geology databases are integrated with facility or site maps to prepare data visualization maps, time series charts, and geologic sections. *Site GIS* interfaces with both Golden Software graphics software and Microsoft *Excel* and is an application for a leading desktop mapping package, *MapInfo* (MapInfo Corp., 200 Broadway, Troy, NY 12180).

Geologic, hydrogeologic, and geochemical data can be analyzed and presented with ***GEOBASE,*** distributed by Earthware of California (30100 Town Center Drive, #196, Laguna Niguel, CA 92677) or with one of several software packages listed in the catalog of the Scientific Software Group (P.O. Box 23041, Washington, D.C. 20026–3041).

Model design database sources include conventional database software such as geographic information systems (GIS), which organize, store, and output hydrogeologic information (see ESRI, 1993). Rarely can data be transferred directly from a GIS or other conventional database into model design software without some kind of data manipulation. As a result, modelers have to create and utilize their own interface software (see Rifai et al., 1993).

DesignMod (Design Model), included with this book, is an inexpensive *Windows* groundwater modeling utility application developed using *Visual Basic for Windows Professional Edition,* version 2.0 and sets of custom controls called *3-D Widgets. DesignMod* executable and exercise (but not source code) files are provided on a disk distributed with this book. *DesignMod* provides a fast, flexible, versatile, and interactive *Windows* graphical user interface method for designing a finite-difference groundwater flow or transport model. *DesignMod* does not directly interface with GIS software but is compatible with conventional databases. It complements the applied groundwater modeling book written by Anderson and Woessner (1992) and enables the user to graphically visualize groundwater model design in relation to geologic, hydrogeologic, and geochemical cross sections and plan view maps.

DesignMod enables the user to design a groundwater model more efficiently and effectively than is possible using the traditional system of paper cross sections, maps, and acetate overlays. Although *DesignMod* is not a preprocessor for any specific groundwater model, it can be used to create and edit some groundwater model data and flag arrays. *DesignMod* quickly and easily processes and manipulates groundwater modeling array data, thereby supplementing groundwater model pre- and postprocessors. Array data can represent hydrostratigraphic unit tops, bases, or thicknesses; active, inactive, or constant head cells; hydraulic conductivity, thermal conductivity, storativity, porosity, or leakance; recharge or evapotranspiration; head or fluid pressure; solute concentration; fluid temperature; etc.

With *DesignMod* you can create, edit, and view a file catalog; browse ASCII data files; load and save log, grid spacing, doublet, scattered triplet, gridded triplet, and grid data array files; convert data array values from one system of units to another system of units; develop and edit model well log databases; create and edit blocks (groups of groundwater model grid cells) of grid spacing, scattered triplet, and Z or XYZ data arrays using replace, addition, subtraction, multiplication, division, or conditional search operators; calculate surface polygon areas; control data means (arithmetic, harmonic, geometric, vertical, and horizontal), equivalent sources and sinks, and critical time step values; interpolate between points on surfaces using triangulation or kriging techniques; view and print grid, triplet, and doublet data array facts; view and print tables of well log data, grid spacing data arrays, grid data arrays, doublet data arrays, scattered triplet data arrays, and gridded triplet data arrays; view and print Z value zonal diagrams and Z value contour maps; view and print arithmetic, semilogarithmic, or logarithmic XY graphs; and draw geologic, hydrogeologic, and geochemical cross sections and plan view maps in the Design Window (drawing pad).

DesignMod supports only finite-difference rectangular grids and defines the groundwater model grid using either the column and row or X- and Y-coordinate matrix. Model grid column numbers and X-coordinates increase from left to right, and row numbers and Y-coordinates increase from top to bottom (upper left origin). Site base map coordinates use the Cartesian notation; X-coordinates increase left to right and Y-coordinates increase bottom to top (lower left origin). Groundwater model cells are centered at nodes, and cell borders are delineated by gridlines halfway between nodes. Both uniform and variable model grid spacings (cell widths) are supported. Array and subarray (block) dimensions are defined by array upper left and lower right corner column and row numbers and/or upper left corner X- and Y-coordinates and column and row grid spacings.

DesignMod supports the following groundwater model data file types: log, grid spacing, doublet, scattered triplet, gridded triplet, and grid. A log file is an ASCII text, sequentially formatted file containing well log or model layer data. Each line has several items of well or model data (numeric and string), separated by commas, and the same type of data after each comma. Data for the smallest well, model, unit, layer, and constituent numbers are on the first line, and data for the largest well, model, unit, layer, and constituent numbers are on the last line. A grid spacing file is an ASCII text, sequentially formatted file containing a two-dimensional (2-D) array of model column or row numbers and associated grid spacings. Each line has one column or row number and one grid spacing value, separated by a comma. The grid spacing for the smallest column or row number is on the first line, and the grid spacing for the largest column or row number is on the last line. A doublet file is an ASCII text, sequentially formatted file containing a 2-D array of X- and Y-coordinates. Each line has one X-coordinate value and one Y-coordinate value, separated by a comma. The Y-coordinate for the smallest X-coordinate is on the first line, and the Y-coordinate for the largest X-coordinate is on the last line.

A scattered triplet file is an ASCII text, sequentially formatted file containing a three-dimensional (3-D) array of X-, Y-, and Z-coordinates. Each line has one X-coordinate value, one Y-coordinate value, and one Z-coordinate value, separated by commas. The file has the following line-by-line format: $1,1,Z(1,1);2,1,Z(2,1); \ldots$ up to $NX,1,Z(NX,1);$ then $1,2,Z(1,2); 2,2,Z(2,2); \ldots$ up to $NX,2,Z(NX,2)$. These entries continue through $NX,NY,Z(NX,NY)$, where NX is the largest X-coordinate and NY is the largest Y-coordinate. The X- and Y-coordinate origin can be either model upper left or site base map lower left. A gridded triplet file is an ASCII text, sequentially formatted file containing a 3-D array of X-, Y-, and Z-coordinates. Each line has one X-coordinate value, one Y-coordinate value, and one Z-coordinate value, separated by commas. The file has the following line-by-line format: $1,1,Z(1,1);2,1,Z(2,1); \ldots$ up to $NX,1,Z(NX,1);$ then $1,2,Z(1,2); 2,2,Z(2,2); \ldots$ up to $NX,2,Z(NX,2)$. These entries continue through $NX,NY,Z(NX,NY)$, where NX is the largest X-coordinate and NY is the largest Y-coordinate. The X- and Y-coordinate origin is model upper left.

A grid file can be either an ASCII text, sequentially formatted or a binary sequentially unformatted file containing a 2-D array of Z values. An ASCII grid file can have a comma-separated value (CSV) or listed space-separated value (LSSV) format. In the CSV file, there are several Z values on a line depending on the value format and the total number of Z values. All Z values except the last are followed by a comma. In the LSSV file, there are six or less Z values with a specified exponential format on a line depending on the total number of Z values. Each Z value on a line (except the last Z value on that line) is followed by a space. The CSV and LSSV files have the following row by row format: $Z(1,1);Z(2,1);$ … up to $Z(NC,1)$, where NC is the number of cells along the model column direction; then $Z(1,2);Z(2,2);$ … up to $Z(NC,2)$. These entries continue through $Z(NC,NR)$, where NR is the number of cells along the row direction. A binary file has no carriage return separating individual Z values. The file is in the form of numeric storage unit bit patterns which are implementation-dependent. ASCII text files can be read, but binary files cannot be read, with the TYPE DOS command.

DesignMod supports any consistent system of units but cannot detect the use of inconsistent units.

2   MODEL DESIGN GUIDELINES

INTRODUCTION

An entire groundwater modeling protocol is described in detail by Anderson and Woessner (1992, pp. 6–9). This chapter provides guidelines for implementing with *DesignMod* the following parts of that protocol: conceptual model development and model design. It is assumed that the reader has a working knowledge of physical and chemical hydrogeology as described by Domenico and Schwartz (1990), is acquainted with modeling theory as described by Kinzelbach (1986), and is fully aware of the limitations of modeling as described by the Committee on Ground Water Modeling Assessment (CGWMA, 1990).

After the purpose of the model has been established, the next step is to develop a conceptual model (set of assumptions and concepts) of groundwater conditions patterned after simple prototype aquifer systems, initial flow and transport conditions, boundaries, sources, and sinks. A conceptual model simplifies field conditions to make them fit an existing mathematical model and is usually displayed in the form of a block diagram or a cross section.

Conceptual models are based on a semiqualitative analysis of data concerning the physical and chemical characteristics and the lateral and vertical dimensions of geologic formations (aquifer system framework and boundaries), chemical and temperature characteristics of groundwater, and the geometry, strength, chemistry, and temperature of sources and sinks. A synthesis is made of available well logs and other hydrogeologic data so that formations can be classified as aquifer, confining bed (aquitard), or aquiclude layers using the concept of the hydrostratigraphic unit (Maxey, 1964; Seaber, 1988).

The occurrence of groundwater is classified as confined, confined/unconfined, or unconfined, and the thickness and boundaries of layers, hydraulic characteristics (horizontal and vertical hydraulic conductivity, storativity, specific yield, porosity, bulk density, dispersivity, etc.) of layers, initial heads in layers, chemical concentration of groundwater in layers, temperature of groundwater in layers, and possible chemical reactions in layers are identified and quantified in general terms. Data can come from pumping, slug, tracer, laboratory, and other tests; surveys; inquiries; published and unpublished databases; literature; and indirect correlation information.

The conceptual model is refined by identifying the plan view area of interest; selecting the model grid location, orientation, and borders; selecting the vertical depth of interest and number of layers in the model; setting the model areal boundary and initial conditions; selecting range values for model layer hydraulic characteristics and hydrologic stresses based on hydrogeologic data, logic, and judgment; and selecting model grid spacings and time steps. Specific values for model layer hydraulic characteristics and hydrologic stresses within range values are established later in the modeling protocol with model calibration data, logic, and judgment. Too frequently, range value selection is slighted or neglected. This practice can lead to incorrect model predictions based primarily on unbounded calibration analyses.

The plan view area of interest, which may or may not encompass the natural boundaries of the aquifer system, is defined so that the model boundaries and initial conditions can be set. The area of interest should cover at least the impact area of pertinent sources and sinks. The head or flux must be specified along the borders of the area of interest. The vertical depth of interest is defined so that the number of layers (hydrogeologic units) can be set. The vertical depth of interest should cover at least the impact depth of pertinent sinks and sources. The lateral and vertical dimensions of the model and flow conditions at the borders of the model are approximated with boundaries (see Franke et al., 1987; Franke and Reilly, 1987).

Model design is an iterative process and includes the careful study of geologic, hydrogeologic, and geochemical cross sections and plan view surface maps. First trial model grid location, orientation, and borders are determined using the site base map and information gained from drawing cross sections and plan view maps. Grid orientation refers to the rotation of the model grid relative to the site base map coordinate system and the model grid origin relative to the site base map origin. The first trial grid is refined by comparing the grid with Z value surface maps (contour and zonation) and the site base map. It is common practice to draw cross sections on graph paper with colored pencils and drafting tools and to overlay the site base map with Z value surface maps and trial grids drawn on tracing paper or clear acetate. *DesignMod* automates this process, thereby easing iterative model design by enabling the user to create and modify cross sections and plan view maps interactively on the screen.

GRID ORIENTATION AND SPACING

Grid orientation is selected in context with the area of interest, the conceptual model, sinks, sources, and points of interest. Grid borders are usually oriented so that they are parallel to the principle directions of the transmissivity tensor (Trescott et al., 1976, p. 30) and most cell nodes will be active. Grid borders are often aligned with flow direction and/or boundaries, and they are centered as closely as possible on sources, sinks (extraction and injection wells, recharge wells, drains, mines, streams, and lakes), and monitoring wells. Transport model grid border orientation should be selected with the understanding that contaminant plumes tend to be elliptical in shape and slug sources produce detached contaminant plumes.

Model grid borders sometimes represent lateral impermeable (barrier) boundaries. Model grids can be expanded beyond the area of interest, and grid borders can be located at sufficient distance so they do not affect results in the area of interest during the simulation time. The grid can be restrained to cover only the area of interest by specifying flows into or from the area of interest. These fluxes are apportioned evenly to cells along appropriate grid borders of the area of interest as prescribed flux boundaries.

In selecting the model grid spacings, a rectangular grid is superposed over a plan view area of interest, and the aquifer system beneath this area is subdivided into discrete volumes (cells). The grid spacing is kept small in comparison to the area of interest to ensure high-precision calculations. Uniform (square) grid spacing is desirable; however, variable grid spacing is often necessary to minimize the number of columns and rows. Pure flow-model grid spacings are usually much larger than those of pure transport models. Commonly, telescopic grid spacing techniques (Ward et al., 1987) are required in flow-transport modeling so that regional, local, and site dimensions are considered. These techniques involve creating separate models for regional, local, and site scales and linking the models by using output from a larger-scale model to simulate initial, boundary, and characteristic conditions in the next smaller-scale model.

In variable grid spacing, the smallest model grid spacings are specified where gradients vary sharply in the vicinity of sources and sinks and where there are rapid changes in aquifer hydraulic characteristics. Grid spacings are gradually increased outward from sources and sinks. It is frequently desirable to specify variable grid spacings with large grid spacings along model borders. The ratio of adjacent grid spacings is usually less than five except for high-precision calculations, when it is two or less (Trescott et al., 1976, p. 30). Grid spacings are uniform along individual columns and rows and are the same in each model layer of a multilayered aquifer system. To minimize numerical dispersion in transport models, the grid should be selected so that a representative grid spacing is less than four times a representative dispersivity, and the time step should be less than a representative grid spacing divided by a representative flow velocity (see Anderson and Woessner, 1992, p. 327).

Digitized site base maps containing objects such as streams, lakes, roads, buildings, well locations, waste disposal sites, etc. are used as background information in selecting grid orientation and spacings. Grid orientation and spacings are varied until the grid closely fits points and conditions of interest.

Commonly, a piece of transparent material is superimposed over a map of the area of interest, and grid lines are drawn and erased until they meet project objectives. *DesignMod* enables you to quickly overlay and erase grids on a site base map.

SITE BASE MAP FEATURE SIMULATION

Model design is aimed at producing a database (quantitative expression of the conceptual model) consisting of grid nodal variable value arrays, selected grid nodal variable value listings, and model-wide variable value listings. Model databases are developed using principles of discretization, aggregation, averaging, and approximation. The most difficult parts of the model database to develop are the grid nodal value arrays.

In designing a model, irregular and/or discontinuous aquifer system boundaries are discretized in a rectangular cell step fashion. Discretization in the vertical direction is accomplished by specifying the number of layers (aquifers and/or aquifer permeable zones). Vertical discretization is governed in part by the vertical resolution desired. Layers can be discontinuous and can represent intervals within which vertical head loss is negligible. A confining bed may be simulated as a layer with low horizontal and vertical hydraulic conductivity and moderate storativity.

Model simulation time is discretized into stress periods during which all source and/or sink stresses are assumed to be constant. Stress periods are divided into nonuniform time steps that increase in length in a stepwise fashion as a geometric progression of a time step multiplier between 1 and 1.5. Time steps are kept small in comparison to the length of the simulation time. In general, the smaller the time step, the greater the precision of source and/or sink impact calculations. The initial time step (critical time step) should be small enough to allow for the explicit formulation of the governing equation (see Anderson and Woessner, 1992, p. 205).

Continuous variable source recharge and/or sink discharge rates are discretized in a rectangular rate step fashion. Each rectangular rate step represents a different stress period with a uniform rate, chemical concentration, and temperature. Commonly, the discharges or recharges from closely spaced sinks or sources are aggregated (lumped) and assigned to an equivalent single sink or source. Several permeable zones are often aggregated into an equivalent single layer. Heads, chemical concentrations, or temperatures vary from one depth interval to another where there are vertical gradients. Two or more depth intervals are aggregated into an equivalent single layer in designing some two-dimensional models. Two-dimensional simulation of the three-dimensional phenomena are least accurate in the immediate vicinity of sinks and sources. Model design is frequently based on the mean or average value of a set of hydraulic characteristic data. Within certain zones of an aquifer system, hydraulic characteristics are relatively homogeneous and average hydraulic characteristic values are meaningful.

Anisotropy can be simulated as part of model design by designating two values of hydraulic conductivity within cell blocks, one along rows and one along columns. Hydraulic conductivity along rows is specified, and the hydraulic conductivity along columns is calculated as the product of the hydraulic conductivity along rows and an anisotropy factor that is uniform throughout the entire grid.

Uniform initial heads, chemical concentrations, and temperatures are specified in model design to simulate no flow and no spatial changes in chemical concentration and temperature in a conceptual model at the start of the simulation time. Variable initial heads, chemical concentrations, and temperatures are specified to simulate flow and spatial changes in chemical and temperature conditions in a conceptual model at the start of the simulation time. If initial heads, chemical concentrations, and temperatures vary in space at the start of the simulation time, then heads, chemical concentrations, and temperatures will change during the simulation period not only in response to specified stresses but also due to the prescribed initial conditions.

Steady-state initial head, chemical concentration, and temperature conditions are estimated in model design by omitting man-made sources and sinks and simulating only the natural flow system with appropriate boundaries, sources, and sinks (Franke et al., 1987). Initial regional flow in model design can be approximated with two parallel constant head boundaries or with two parallel lines of closely spaced wells. The prescribed heads at the constant head boundaries or the well discharge and recharge rates differ in such a manner as to create a designed regional hydraulic gradient with an observed flow rate.

Impermeable boundaries are simulated in model design as inactive (no flow) cells. Usually, special head values are specified for inactive cells to provide identification during output data processing. In plan view, no-flow boundaries are often aligned parallel to the direction of flow and oriented parallel to flow lines (streamlines) such as topographic highs and/or lows, thereby assuming that these flow lines (groundwater divides) will not change as a result of stresses. Parallel flow line boundaries are often specified to reduce the size of the grid. Sometimes a saltwater/freshwater interface is simulated in model design as a no-flow boundary, thereby assuming that saltwater and freshwater are immiscible.

Variable recharge (infinite source such as a surface water body) and/or discharge (infinite sink such as a surface water body) boundaries are simulated in model design as constant value cells with specified heads, concentrations, and temperatures. In cross section, an upper boundary representing the water table is approximated as a prescribed flux (Neumann) condition by specifying recharge and/or discharge (evapotranspiration) rates at each node along the boundary. The base of a conceptual model is commonly simulated as a no-flow boundary.

Regional flow is frequently simulated in model design with two parallel constant head boundaries (upgradient and downgradient) having different head, concentration, or temperature values. Constant head boundaries are specified where recharge or underflow into or out of an impact area is sufficient to maintain a nearly constant head at that point in the conceptual model.

Constant specified flux (flow rate of water, contaminant mass, and/or energy) boundaries are simulated in model design with closely spaced adjacent recharge and/or discharge well cells. Semipervious boundaries are simulated with general-head cells, which are similar to river cells in that flow is provided in proportion to constant external source heads and variable cell heads.

Slug sources are simulated in model design as initial contaminant concentrations or as short-term contaminant flux rates. Continuous contaminant sources are simulated as injection cells with prescribed recharge rates and contaminant concentrations or as cell leakances with prescribed leakance coefficients and contaminant concentrations. Boutwell et al. (1986, pp. 106–108) describes several methods for estimating waste disposal facility source seepage rates. Glass (1989, pp. 1068–1069) describes a method for simulating a source whose strength decreases as the contaminant is leached from the soil.

A multiaquifer production or injection well can be simulated in model design by assigning unique discharge or injection rates to the same node in each aquifer open to the multilayer well. The rates are calculated as the product of the total rate and the particular aquifer transmissivity divided by the sum of the aquifer transmissivities.

Flowing wells and springs may be simulated in model design as drainage leakage nodes with leakances proportional to the effective radius of the well or spring (Walton, 1991, pp. 315–317). Ditches, mines, and tunnels may be simulated as constant head nodes with mine node transmissivity adjustment factors (Walton, 1991, pp. 319–320) or as stream drainage with stages below the water table or potentiometric surface. Ditches, mines, and tunnels may also be simulated as drains by setting the drain nodal length equal to the ditch, mine, or tunnel nodal length with the drain nodal diameter equal to the ditch, mine, or tunnel nodal width, the drain nodal depth equal to the ditch, mine, or tunnel nodal depth, and the thickness and vertical hydraulic conductivity of the drain nodal fill equal to the thickness and vertical hydraulic conductivity of the ditch, mine, or tunnel nodal bed. Gravity drainage of the mine or tunnel cavity is usually calculated external to a model and added to the mine or tunnel drainage rate calculated with a model. Mine drift stepwise advancement may be simulated by extending the mine sequentially in time to adjacent nodes.

Narrow structures such as fault zones or slurry walls can be simulated in model design by using extremely small finite-difference cells, or the horizontal conductance between adjacent cells can be replaced by a much smaller intercell conductance computed for a thin barrier of low hydraulic conductivity. One of the four horizontal faces of the cell contains the wall (Hsieh and Freckleton, 1992).

3   FEATURES OF COMMERCIAL *WINDOWS* SOFTWARE

INTRODUCTION

DesignMod is compatible and can communicate with most commercial word processor, spreadsheet, database management system, and *Windows* graphics software through the *Windows* clipboard and ASCII text and bitmap files. Word processors, spreadsheets, database managers, and graphics programs can import, edit, and display *DesignMod* text and graphics files and share the *Windows* clipboard with *DesignMod*. *DesignMod* can import, edit, and display text and graphics files created with word processors, spreadsheets, and database managers that support several sophisticated editing, calculating, and viewing tools that complement the custom tools supported by *DesignMod*.

WORD PROCESSOR

You can open *DesignMod* files directly in Microsoft's *Word for Windows*, version 6.0 ($300), word processor software. *Word* recognizes *DesignMod* file formats (text and bitmap) and converts *DesignMod* files into *Word* format, preserving the content and formatting of the *DesignMod* files. You can convert *Word* format documents directly into *DesignMod* formats when you save the document. You also can convert *DesignMod* files imported into *Word* to many file formats by using *Word* converters. *Word* provides conversions for the following file formats:

Word for Windows 1.x and 2.x; *Word* for DOS 4.0, 5.0, and 5.5; and *Word* for the Macintosh 4.0 and 5.0.

WordPerfect 4.1, 4.2, 5.0, 5.1, and 6.0

RFT-DCA (for *DisplayWrite* and IBM 5520)

Works for Windows and *Works* 3.0 for DOS (word processing documents only)

WordStar 3.3, 3.45, 4.0, 5.0, and 5.5
Lotus 1-2-3, 2.x and 3.0 (convert to *Word* format only, not from *Word* format)
Microsoft *Excel* BIFF 2.x and 3.0 (convert to *Word* format only)
Multiplan 3.0 and 4.2
dBASE (II, III, II Plus, and IV)
Text (with options for text only, text with line breaks, and text with layout)
DOS text (with options for DOS text only, DOS text only with line breaks, and DOS text with layout)
Rich Text Format (RTF)

DesignMod text and graphics files and clipboard contents can be incorporated into documents such as reports in *Word* using its convenient Menu bar, Toolbar, Ribbon, Ruler, and Document window as well as many basic word processing features including moving and copying text using the clipboard, moving text on a page by dragging it with a mouse, and multicolumn formatting. *DesignMod* array files can be incorporated easily into model databases with *Word*'s extensive editing tools.

Word supports object linking and embedding (OLE). You can add charts, including arithmetic, semilogarithmic, and logarithmic XY graphs, to a document using Microsoft *Graph,* the embedded charting application that comes with *Word*. Charts can be sized and extensively edited in a chart window. You can create your own graphics and add them to a document using Microsoft *Draw,* the embedded drawing application that comes with *Word*. Microsoft *Draw* has the standard palette of drawing tools that makes it simple to move and change the size, shape, and color of images and supports flipping, rotating, and zooming in and out. With Microsoft *Draw,* you can import graphics into a document in the following graphic formats:

Windows Metafile (*.WMF)
Encapsulated PostScript (*.EPS)
TIFF (tagged image file format) (*.TIF)
Computer Graphics Metafile (*.CGM)
HP Graphic Language (*.HGL)
DrawPerfect (*.WPG)
Micrografx Designer 3.0/*Draw Plus* (*.DRW)
PC Paintbrush (*.PCX)
Windows bitmaps (*.BMP)
AutoCAD 2-D format (*.DXF)
AutoCAD Plotter format (*.PLT)
Lotus 1-2-3 Graphics (*.PIC)

SPREADSHEET

You can open *DesignMod* files directly in Microsoft's *Excel for Windows*, version 5.0 ($300), spreadsheet software. *Excel* recognizes *DesignMod* file formats (text and bitmap) and converts *DesignMod* files into *Excel* format, preserving the content and formatting of the *DesignMod* files. You can convert *Excel* format documents directly into *DesignMod* formats when you save the document. You can export flat files in which data are separated by space characters instead of tabs or commas. *Excel* supports dynamic data exchange (DDE) and OLE. You can convert *DesignMod* files imported into *Excel* to many file formats by using *Excel* converters. *Excel* provides conversions for the following file formats:

Format	Type of Document
Normal	Standard Microsoft *Excel*, version 5.0 format
Template	Microsoft *Excel*, version 5.0 template format
Excel 4.0	Microsoft *Excel*, version 4.0
Excel 3.0	Microsoft *Excel*, version 3.0
Excel 2.x	Microsoft *Excel for Windows*, version 2.1; Microsoft *Excel* for the Macintosh, version 2.2; or Microsoft *Excel* for OS/2, version 2.2
SYLK	Symbolic link format (Microsoft *Multiplan*)
Text	ANSI text for *Windows*, text for the Macintosh or MS-DOS
CSV	Comma-separated values, ANSI text for *Windows*, text for the Macintosh or MS-DOS
WKS	*Lotus 1-2-3*, release 1A
WK1	*Lotus 1-2-3*, release 2.x
FMT	Formatting information for the WYSIWYG add-in for *Lotus 1-2-3*, release 2.x (saved or opened with a WK1 file)
WK3	*Lotus 1-2-3*, release 3.x, and *Lotus 1-2-3 for Windows*
FM3	Formatting information for *Lotus 1-2-3 for Windows* and for the WYSIWYG add-in for *Lotus 1-2-3*, release 3.1 (saved or opened with a WK3 file version 4.0)
DIF	Data interchange format (*VisiCalc*)
DBF2	dBASE II
DBF3	dBASE III
DBF4	dBASE IV

Excel supports the following clipboard formats:

Format	Clipboard Type Identifier
Picture (metafile)	*Picture for Windows;* PICT for the Macintosh
Bitmap	Bitmap for *Windows*
Microsoft *Excel* file formats	BIFF, BIFF3, BIFF4
Symbolic link format (Microsoft *Multiplan*)	SYLK
Lotus 1-2-3, release 2.x format	WK1
Data interchange format (*VisiCalc*)	DIF
Text (tab delimited)	Text
Comma-separated values	CSV
Formatted text	Rich Text Format for *Windows;* RTF for Macintosh
Embedded object	*Native, OwnerLink, Picture for Windows;* NATV, OLNK, PICT for the Macintosh; or other presentation format
Linked object	*OwnerLink, ObjectLink, Link, Picture for Windows;* OLNK, OJLK, LINK, PICT for the Macintosh; or other presentation format
Text	Display Text, OEM Text
Values	VALU for the Macintosh

Excel enables you to store, manipulate, calculate, and analyze array data on a worksheet (spreadsheet); sort, search, and manage array data using standard database operations; present array data in a chart; and add graphic elements to a worksheet using drawing tools. A worksheet is a rectangular grid of up to 256 columns and 16,384 rows. The intersection of each column and row is a cell (corresponding to a groundwater model grid cell) in which you store data. You can store and edit data in a single cell, a range of adjacent cells, or nonadjacent cells and insert or delete cells, rows, and columns. You can create and analyze a database on a worksheet.

Q+E is a database application which comes with *Excel* that enables you to manipulate and update database files from a variety of database systems, including dBASE, Microsoft SQL Server, *Oracle*, and OS/2 Extended Edition. You can analyze cell data with formulas that perform operations such as addition, multiplication, and comparison. Formulas can include operators, cell references, values, and built-in worksheet functions. *Excel* provides a wide variety of functions, including many which perform mathematical actions on values, statistical actions on values, and engineering calculations on values.

A set of special statistical and engineering cell data analysis tools called the *Analysis ToolPak* comes with *Excel* and contains the following tools: Anova: Single Factor, Anova: Two-Factor With Replication, Anova: Two-Factor Without Replication, Correlation, Covariance, Descriptive Statistics, Exponential Smoothing; F-Text: Two-Sample for Variances, Histogram, Moving Average, Random Number Generation, Rank and Percentile, Regression; t-Text: Paired Two-Sample for Means; t-Text: Two-Sample Assuming Equal Variances; and t-Text: Two-Sample for Means, Fourier Analysis, and Sampling.

Excel enables you to create 14 types of charts with worksheet data and the following tools: Area Chart, Bar Chart, Column Chart, Stacked Column Chart, Line Chart, Pie Chart, XY Scattered Chart, 3-D Area Chart, 3-D Bar Chart, 3-D Column Chart, 3-D Perspective Column Chart, 3-D Line Chart, 3-D Pie Chart, 3-D Surface Chart, Radar Chart, Line/Column Chart, Volume/Hi-Lo-Close Chart, Preferred Chart, ChartWizard, Horizontal Gridlines, Legend, Arrow, and Text Box. The 3-D Surface Chart type includes a 3-D surface chart, 3-D wireframe chart, 2-D color contour chart, and 2-D wireframe contour chart.

Excel can be used as a front end for Fortran (Ribar, 1993, pp. 181-203). *Excel* can handle all user interaction and printing and can call Fortran functions embedded in dynamic link libraries to perform calculations. *Excel* provides tools for creating professional user interfaces, including features that handle screen design and printer support. *Excel* has the capability of adding programming functions to the spreadsheet. *Excel* 5.0 contains the *Visual Basic Applications Edition*, a powerful programming language that comes with a complete set of tools for developing custom applications and macros. Within 2 years, *Word* and *Access* will come with *Visual Basic*.

DATABASE MANAGEMENT

You can open *DesignMod* files such as well log files directly in Microsoft's *Access for Windows*, version 1.1 ($319), relational database management system. *Access* recognizes *DesignMod* file formats (text and bitmap) and converts *DesignMod* files into *Access* format, preserving the content and formatting of the *DesignMod* files. You can convert *Access* format documents directly into *DesignMod* formats when you save the document. *Access* supports DDE and OLE. You can convert *DesignMod* files imported into *Access* to many file formats by using *Access* converters. *Access* can import data from the following database formats and applications: *Paradox,* dBASE III, dBASE IV, *Btrieve* (with *Xtrieve* dictionary file), Microsoft *SQL* Server, *Oracle* Server, Microsoft *FoxPro* versions 2.0 and 2.5, and Microsoft *Access* (databases other than the open database). *Access* can import or export files in the following spreadsheet and text formats: Microsoft *Excel* (versions 2.x, 3.0, and 4.0), *Lotus 1-2-3* or 1-2-3/W (WKS, WK1, and WK3 files), delimited text (values separated by commas, tabs, or other characters), and fixed-width text (values arrranged so that each field has a certain width). A maximum database file size of 1 gigabyte is supported.

Access enables you to open a database form and enter data in a table, set up and run a database query, build and use a custom database form, and design and print a database report. *Access* automatically maintains database table relationships, performs data validation, and handles printing details without programming. *Access* creates one file containing all the tables in a database and the queries, forms, reports, and other objects that help you use the information in the database. In addition, you can write custom functions for expressions, automate the manipulation of objects and data in your database, and build sophisticated database applications using the *Access* Basic programming language derived from Microsoft's *Visual Basic* language.

Design and datasheet views can be used to create, edit, and view a table. You can create a data entry form with FormWizard and customize data entry forms using a Control Toolbox. You can create select queries to retrieve data from tables and action queries to create new tables or change data in existing tables, sort records, calculate totals, find a value, and create a filter to find records. You can create indexes for tables and use expressions to do string and mathematical calculations. You can graph data from your database tables (including arithmetic, semilogarithmic, and logarithmic XY graphs) using GraphWizard and Graph Tools and create database reports with ReportWizard.

The Microsoft *Access* Distribution Kit (ADK; $495) allows developers to package and distribute their *Access* applications without requiring their users to purchase the full Microsoft *Access* product. The *Access* runtime engine can be distributed royalty-free. The ADK includes the Microsoft *Windows* Help compiler, Wizards that assist developers in creating disk sets, and a custom setup program.

GRAPHICS

3-D Visions Standford Graphics for Windows, version 2.1 ($326), software recognizes *DesignMod* file formats (text and bitmap) and converts *DesignMod* files into *Standford Graphics* formats. You can convert a *Standford Graphics* ASCII data spaced text file with a header into a *DesignMod* text file using a word processor. You can convert *Standford Graphics* images directly into the *DesignMod* image bitmap format when you save the image.

Standford Graphics imports the following data files: *Standford Graphics Presentation* file SG1, *Standford Graphics Slide Show* file SHW, *Standford Graphics* ASCII space-delimited XYZ triplet spreadsheet format file DAT with a header, standard delimited ASCII text XYZ triplet spreadsheet format file TXT, data interchange format file DIF, *Lotus 1-2-3* spreadsheet format files WK1 and WK3, Microsoft *Excel* spreadsheet format file XLS, and *Graftool Presentation* file DCF. *Standford Graphics* exports the following data files: *Standford Graphics Presentation* file SG1, *Standford Graphics Slide Show* file

SHW, and *Standford Graphics* ASCII space-delimited spreadsheet format file DAT with a header. *Standford Graphics* imports images in the following graphic file formats: *Freelance, Harvard Graphics, Mirage,* and *Pixie* ANSI CGM; *Micrografx Designer, Draw Plus,* and *Graph Plus DRW; Plotters, Graftool,* and *Mathcad* H-P HPGL PLT; *PC Paintbrush* IV bitmap paint programs PCX; *CompuServe* graphics GIF; *Windows* 3.x bitmap BMP; *Lotus 1-2-3* PIC; and TIFF 5.0 TIF. *Standford Graphics* exports images as a bitmap BMP or *Windows* Metafile WMF file. *Standford Graphics* supports DDE.

Standford Graphics is a desktop presentation program with a highly integrated set of presentation development and data analysis tools that enables you to graphically explore, analyze, and chart your 2 D and 3-D data with the following processing tools: on-screen curve fitting, regressions, FFTs, smoothing, interpolation, and contouring; 162 graph types for business, statistical, and technical data, including scatter plots (arithmetic, semilog, and log), contour plots, shadow-contour plots (combines a surface plot with a 2-D contour plot), and surface plots; over 70 statistical and scientific functions, including sine, cosine, tangent (hyperbolic and inverse), vector angle, Kroenecker delta, erf(x), erfc(x), exp(x), natural log, base-10 log, pi, min, max, sum, mean, standard deviation, standard error, square root, polarity, random variable, step, sort, Gaussian-Q, gamma, ANOVA, J(n,x), Y(n,x), chi-square, and student's t-tests; automatic processing for contours; built-in 4-D 70-trillion-cell spreadsheet; graphical "what-if" data and formula manipulation; and probability, log, linear, and polar scaling.

You can create speaker notes, audience handouts, and output to any printer supported by *Windows* with *Standford Graphics.* The following types of analyses are supported: statistical-box-whisker, error-bar, histogram, and Pareto; fast Fourier transforms; curve cubic-spline interpolation; 3-D gridding interpolation and contouring; and linear and nonlinear regressions, including cyclical, exponential, geometric, hyperbolic, polynomial, and 3-D surface regressions.

Standford Graphics uses an inverse-distance weighting algorithm to interpolate the 3-D surface grid or matrix required to create contour maps and 3-D surface plots from scattered or regular XYZ triplet data. A rectangular M uniformly spaced by N uniformly spaced interpolation grid is supported. The weighting factor for each grid point Z value is the reciprocal of the distance of the grid point from the control data point raised to some power P. The smoothness of the surface that is generated can be influenced by changing the power P.

4 *VISUAL BASIC* PROGRAMMING TECHNIQUES

INTRODUCTION

Visual Basic enables you to create windows (forms), draw objects (controls) such as text boxes and command buttons on the forms, and program how these forms and controls respond to user actions. *Visual Basic* uses the simplified syntax of BASICA and GW-BASIC and supports nearly all of their capabilities. *Visual Basic* uses the programming structure of QuickBASIC, so you should be thoroughly familiar with that language before developing your first application (see Halvorson and Rygmyr, 1991; Arnson, Gemmell, and Henderson, 1991). Two excellent *Visual Basic* primers are those of Orvis (1992) and Cornell (1992). Guidelines for developing *Visual Basic* applications appear in *The Windows Interface — An Application Design Guide* (Microsoft Press, 1992). The source code for a *DesignMod* form presented in Appendix A illustrates several common *Visual Basic* programming techniques.

There are a large number of books available that supplement the *Visual Basic Programmer's Guide* and can help you learn *Visual Basic* programming techniques. Each book has some material that the other books do not have and is written from a slightly different perspective than the other books. The quick way to learn *Visual Basic* is to read or scan several books, thereby gaining the benefit of individual author perspectives and specialties.

The following sources emphasize subjects not covered in *Visual Basic's Programmer's Guide:*

"Mouse-Driven Paint Program" (Holzner, 1991, pp. 255–306)
"Shell Sorts" (Holzner, 1991, pp. 329–339)
"Numeric Built-In Functions" (Cornell, 1992, pp. 238–241)
"Screen Scales" (Cornell, 1992, pp. 386–392)
"Restricting Data-Entry" (Murray and Pappas, 1992, p. 142)
"Data Entry" (Murray and Pappas, 1992, pp. 213–218)
"Accessing *Windows* GDI Graphics Primitives" (Murray and Pappas, 1992, pp. 242–246)

"Manipulation of Fonts" (Murray and Pappas, 1992, pp. 346–363)
"Open File Form" (Craig, 1991, pp. 55–64)
"Save File Form" (Craig, 1991, pp. 65–71)
"File View Form" (Craig, 1991, pp. 72–76)
"Percent Completed Form" (Craig, 1991, pp. 85–89)
"Set Text Box to Greater than 32K" (Entsminger, 1992, pp. 372–373)
"Dynamic Data Exchange" (Entsminger, 1992, pp. 265–318)
"Object Linking and Embedding" (Entsminger, 1992, pp. 417–462)
"Custom Databases" (Thomas, Arnson, and Waite, 1993)
"Installation Programs" (Gurewich and Gurewich, 1993, pp. 807–828)

Programming techniques for developing *Windows* applications with *Visual Basic* are contained in the monthly publication entitled *Inside Visual Basic* distributed by The Cobb Group (Customer Relations, 9420 Bunsen Parkway, Suite 300, Louisville, KY 40220). For example, the graphics technique of creating scrollable picture controls is described in the April 1992 (Vol. 2, No. 4) issue. The *Basic Users Group* monthly newsletter distributed by MicroHelp, Inc. (4359 Shallowford Industrial Parkway, Marietta, GA 30066) provides tips, tricks, techniques, compiler bugs, and caveats covering *Visual Basic*. Programming techniques are also presented in the bimonthly magazine *Visual Basic Programmer's Journal* (P.O. Box 58871, Boulder, CO 80321–8871). Microsoft University (10700 Northup Way, Bellevue, WA 98004–1447) offers courses for *Windows* software developers.

Windows Application Programming Interface (API) functions, Third Party Custom Controls, Third Party Dynamic Link Libraries (DLLs), and Third Party *Visual Basic* forms can be used to add specialized functionality to, expand capabilities of, speed up, or enhance *Visual Basic*. Icons for loaded Custom Controls are added to the *Visual Basic* Toolbox. These Custom Controls then can be used in the same manner as *Visual Basic's* own tools. DLLs can be declared within *Visual Basic* code and then called from any *Visual Basic* subroutine or module. DLLs are libraries of procedures distributed with your *Visual Basic* .EXE program that *Visual Basic* applications can link to and use at run time. Most Third Party Custom Controls and DLLs can be distributed without royalties.

Third Party Custom Controls, Dynamic Link Libraries, and *Visual Basic* forms (add-ons) perform the following functions: database management, desktop publishing system tailored for databases, micro-to-mainframe client/server, serial communications, image and scanner processing, network interface, report generation, spell checker, programming management, application testing, 1-, 2-, and 3-D business graph generation, 3-D controls and menus, help file development, file compression, spreadsheet generation, word processing, custom control development, 24-bit color support, custom cursors, code debugging, and 3-D scientific and engineering graphic drawing and rendering. For a complete listing of Third Party add-ons, see the Winter/Spring 1994 *Custom*

Controls and Other Companion Products and Services for Visual Basic for Windows, distributed by Fawcette Technical Publications (280 Second Street, Suite 200, Los Altos, CA 94022–3603; 415–917–7650; $7.00 per copy). Third Party add-ons are advertised in the bimonthly *Visual Basic Programmer's Journal* (a Fawcette Technical Publication; P.O. Box 58872, Boulder, CO 80322–8872; 303–541–0610; $27.95 per year).

WINDOWS APPLICATION PROGRAMMING INTERFACE FUNCTIONS

Over 600 functions are provided by the *Windows* system. These functions are collectively termed the *Windows* Application Programming Interface, or API. Most API functions can be called by *Visual Basic* procedures and can be mixed and matched with regular *Visual Basic* functions. API functions can be used to accomplish tasks that are not build into *Visual Basic*, to access advanced features left out of *Visual Basic*, and to speed up program execution. API functions are contained in *Windows* DLL's. The three main DLL files are KERNEL.EXE (provides system services), GDI.EXE (provides the graphics device interface), and USER.EXE (provides window management). These files are usually found in the directory C:\WINDOWS\SYSTEM.

Complete information about the use of API functions is provided in the official documentation of the Microsoft Development Kit (SDK), which is an essential resource for every serious *Windows* programmer. A six-book series of this documentation, called the *Microsoft Windows 3.1 Programmer's Reference Library* (Vol. 1, *Overview*, $29.95; Vol. 2, *Functions*, $39.95; Vol. 3, *Messages, Structures, Macros*, $29.95; Vol. 4, *Resources*, $29.95; Vol. 5, *Programming Tools*, $29.95; and Vol. 6, *Guide to Programming*, $29.95), is available from Microsoft Press through most bookstores. The author has found Volume 2 (*Functions*) and Volume 6 (*Guide to Programming*) to be most useful. These references use standard C language-oriented programming terminology, which is difficult but not impossible to understand by Basic programmers.

Fortunately, extensive information about the use of API functions using *Visual Basic* programming terminology instead of standard *Windows* programming terminology is available in the books of Appleman (1993), Arnson, Rosen, Waite, and Zuck (1992), Murray and Pappas (1992), and Young (1992). The use of many API functions to extend the capabilities of *Visual Basic* is demonstrated in these books; basic code and diskette source files are also provided.

Because the API functions are not an integral part of *Visual Basic*, you must explicitly declare an API function before calling it. The declare statement gives *Visual Basic* access to the API functions. The declaration can be placed within the declarations section of the module from which it is called, or it can be placed in the global module. After you have declared the API function, it can be called in the same manner as you would a regular *Visual Basic* function or statement, although care must be taken in passing arguments to DLLs (see *Visual Basic Programmer's Guide*, Microsoft Press, 1992, pp. 499–500). The function's parameters and their specific order and type are included in the declaration. Basic information concerning passing arguments to DLLs is given in the book by Appleman (1993, pp. 51, 912–913).

The declarations for API functions are fairly complex. They can be copied from the file WINAPI.TXT in the *Visual Basic* subdirectory \WINAPI. That file contains DLL procedure, constant, and user-defined type declarations for the *Windows* API functions in *Visual Basic* syntax. Declarations can be pasted into *Visual Basic* code using a word processor (see pp. 491–505 in *Microsoft Visual Basic Programming System for Windows*, Version 2.0, Microsoft Press, 1992). In addition, the file WINAPI.HLP in the *Visual Basic* subdirectory \WINAPI contains documentation for the entire *Windows* 3.1 API, and the file WIN31API.HLP in the *Visual Basic* subdirectory \WINAPI contains an indexed and cross-referenced list of API functions. You can view these files by choosing the VB Professional ToolKit icon within Program Manager.

Visual Basic supports the printing of horizontal text; it does not support the printing of vertical text such as graph Y-axis labels. Program instructions and code for manipulating text fonts are herein provided to illustrate how three API functions (CreateFont, SelectObject, and TextOut) are called in *Visual Basic* so that text can be printed at any angle, thereby making it possible to create vertical graph Y-axis labels. CreateFont uses 14 parameters (all of which can be variables) to design a custom font. SelectObject is passed two parameters (the first parameter cannot be a variable; the second parameter can be a variable) when called, and TextOut is passed five parameters (the first parameter cannot be a variable; other parameters can be variables) when called.

According to the *Windows Guide to Programming* (Chapter 18, "Fonts," pp. 18.1–18.4), text fonts can be manipulated by (1) assigning the API CreateFont function return value (integer) to a variable such as CF; (2) assigning the API SelectObject function return value (integer) to another variable such as SO and letting the hDC property (device context handle) of the *Visual Basic* form, picture box, or printer be the first SelectObject function parameter and hnFont be the second parameter; and (3) assigning the API TextOut function return value (integer) to a third variable such as TO and letting the hDC property (device context handle) of the *Visual Basic* form, picture box, or printer be the first TextOut function parameter. The *Windows* environment manages the system display by assigning a device context for the printer object and for each form and picture box in an application (see *Visual Basic Programmer's Guide*, Microsoft Press, 1992, p. 503). You can refer to the handle for an object's device context with the hDC property, which provides a value to pass to *Windows* API calls.

Declarations for the CreateFont, SelectObject, and TextOut functions, respectively, which must be entered on a single unbroken line of program code, are as follows:

```
Declare Function CreateFont Lib "GDI" (ByVal H, ByVal W,
ByVal E, ByVal O, ByVal WT, ByVal I, ByVal U, ByVal S,
ByVal CS, ByVal OP, ByVal CP, ByVal Q, ByVal P, ByVal
Face As String)

Declare Function Select Object Lib "GDI" (ByVal hDC As
Integer, ByVal hObject As Integer)

Declare Function TextOut Lib "GDI" (ByVal hDC As Integer,
ByVal X As Integer, ByVal Y As Integer, ByVal P As String,
ByVal N As Integer)
```

With these declarations and variable proper variable definition, the following example code will print a graph vertical Y-axis label in a picture box Pict1 on form DR15:

```
CF = CreateFont(12, 10, 900, 900, 700, 0, 0, 0, 255, 0, 0, 0, 2,
      "Ms Sans Serif")

SO = SelectObject(DR15Pict1.hDC, CF)

TO = TextOut(DR15Pict1.hDC, 150, 15, "Concentration in
      mg/L")
```

The syntax, parameters, and possible return values for the CreateFont, SelectObject, and TextOut functions follow.

CreateFont

```
Syntax HFONT CreateFont (nHeight, nWidth, nEscapement,
nOrientation, nWeight, cItalic, cUnderline, cStrikeOut,
cCharSet, cOutputPrecision, cClipPrecision, cQuality,
cPitchAndFamily, lpFacename)
```

This function creates a logical font that has the specified characteristics. The logical font can subsequently be selected as the font for any device. Font dimensions are described in pixels.

Parameter	Type/Description
nHeight	int/Specifies the desired height (in logical units) of the font; the font height can be specified in three ways. If nHeight is greater than zero, it is transformed into device units and matched against the cell height of the available fonts. If it is zero, a reasonable default size is used. If it is less than zero, it is transformed into device units and the absolute value is matched against the character height of the available fonts. For all height comparisons, the font mapper looks for the largest font that does not exceed the requested size and, if there is no such font, looks for the smallest font available; 10 is standard.
nWidth	int/Specifies the average width (in logical units) of characters in the font. If nWidth is zero, the aspect ratio of the device will be matched against the digitization aspect ratio of the available fonts to find the closest match, determined by the absolute value of the difference; 8 is standard.
nEscapement	int/Specifies the angle (in tenths of degrees) of each line of text written in the font (relative to the bottom of the page). The angle is measured counter-clockwise from the X-axis. Escapement is character baseline; 0 is horizontal and 900 is vertical. 900/10 = degrees.
nOrientation	int/Specifies the angle (in tenths of degrees) of each character's baseline (relative to the bottom of the page). The angle is measured in a counterclockwise direction from the X-axis when Y-direction is down and in a clockwise direction from the X-axis when Y-direction is up. Orientation is character rotation; 0 is none and 900 is rotated.
nWeight	int/Specifies the desired weight of the font in the range 0 to 1000 (for example, 400 is normal, 700 is bold). If nWeight is zero, a default weight is used. This parameter can be one of the following values: FW_DONTCARE 0, FW_THIN 100, FW_EXTRALIGHT 200, FW_ULTRALIGHT 200, FW_LIGHT 300, FW_NORMAL 400, FW_REGULAR 400, FW_MEDIUM 500, FW-SEMIBOLD 600, FW_DEMIBOLD 600, FW_BOLD 700, FW_EXTRABOLD 800, FW-ULTRABOLD 800, FW_BLACK 900,

Parameter	Type/Description
	FW_HEAVY 900
cItalic	BYTE/Specifies whether the font is italic.
cUnderline	BYTE/Specifies whether the font is underlined.
cStrikeOut	BYTE/Specifies whether characters in the font are struck out.
cCharSet	BYTE/Specifies the desired character set. The following values are predefined:

ANSI_CHARSET 0, DEFAULT_CHARSET 1,
SYMBOL_CHARSET 2, OEM_CHARSET
(standard 255)

The OEM character set is system-dependent. Fonts with other character sets may exist in the system. If an application uses a font with an unknown character set, it should not attempt to translate or interpret strings that are to be rendered with that font. Instead, the strings should be passed directly to the output device driver.

cOutputPrecision	BYTE/Specifies the desired output precision. The output precision defines how closely the output must match the requested font's height, width, character orientation, escapement, and pitch. It can be any one of the following values:

OUT_CHARACTER_PRECIS,
OUT_DEFAULT_PRECIS, OUT_STRING_PRECIS,
OUT_STROKE_PRECIS

cClipPrecision	BYTE/Specifies the desired clipping precision. The clipping precision defines how to clip characters that are partially outside the clipping region. It can be any one of the following values:

CLIP_CHARACTER_PRECIS,
CLIP_DEFAULT_PRECIS, CLIP_STROKE_PRECIS

cQuality	BYTE/Specifies the desired output quality. The output quality defines how carefully GDI must attempt to match the logical-font attributes to those of an actual physical font. It can be any one of the following values:

DEFAULT_QUALITY, DRAFT_QUALITY,
PROOF_QUALITY

Parameter	Type/Description
cPitchAndFamily	BYTE/Specifies the pitch and family of the font. The two low-order bits specify the pitch of the font and can be any one of the following values:

DEFAULT_PITCH, FIXED_PITCH
(1),VARIABLE_PITCH (2).

The four high-order bits of the field specify the font family and can be any one of the following values:

> FF_DECORATIVE (5 < 4, novelty fonts; OLD ENGLISH is an example), FF_DONTCARE (0 < 4), FF_MODERN (3 < 4, fonts with constant stroke width, with or without serifs; Pica, Elite, and Courier New are examples), FF_ROMAN (1 < 4, fonts with variable stroke width and with serifs; Times New Roman and New Century Schoolbook are examples), FF_SCRIPT (4 < 4), FF_SWISS (2 < 4, fonts with variable stroke width and without serifs; MS Sans Serif is an example).

Parameter	Type/Description
lpFacename	LPSTR/Points to a null-terminated character string that specifies the typeface name of the font. The length of this string must not exceed 30 characters. The EnumFonts function can be used to enumerate the typeface names of all currently available fonts.
Return Value	The return value identifies a logical font if the function is successful; otherwise, it is NULL.
Comments	The CreateFont function does not create a new font. It merely selects the closest match from the fonts available in GDI's pool of physical fonts.

SelectObject

> Syntax HANDLE SelectObject(hDC, hObject)

This function selects the logical object specified by the hObject parameter as the selected object of the specified device context. The new object replaces the previous object of the same type. For example, if hObject is the handle to a logical pen, the SelectObject function replaces the selected pen with the pen specified by hObject. Selected objects are the default objects used by the GDI output functions to draw lines, fill interiors, write text, and clip output to specific areas of the device surface. Although a device context can have six selected objects (pen, brush, font, bitmap, region, and logical palette), no more than one object of any given type can be selected at one time. SelectObject does not select a logical palette; to select a logical palette, the application must use SelectPalette.

Parameter	Type/Description
hDC	HDC/Identifies the device context.
hObject	HANDLE/Identifies the object to be selected. It may be any one of the following, and it must have been created by using one of the following functions: Object; Function Bitmap (bitmaps can be selected for memory device contexts only and for only one device context at a time) — CreateBitmap, CreateBitmapIndirect, CreateCompatibleBitmap, CreateDIBitmap; Brush — CreateBrushIndirect, CreateHatchBrush, CreatePatternBrush, CreateSolidBrush; Font — CreateFont, CreateFontIndirect; Pen — CreatePen, CreatePenIndirect; Region — CombineRgn, CreateEllipticRgn, CreateEllipticRgnIndirect, CreatePolygonRgn, Create RectRgn, CreateRectRgnIndirect.
Return Value	The return value identifies the object being replaced by the object specified by the hObject parameter. It is NULL if there is an error. If the hDC parameter specifies a metafile, the return value is nonzero if the function is successful; otherwise, it is zero. If a region is being selected, the return is the same as for SelectClipRgn.
Comments	When you select a font, pen, or brush by using the SelectObject function, GDI allocates space for that object in its data segment. Because data-segment space is limited, you should use the DeleteObject function to delete each drawing object that you no longer need. Before deleting the last of the unneeded drawing objects, an application should select the original (default) object back into the device context. An application cannot select a bitmap into more than one device context at any time.

TextOut

Syntax BOOL TextOut(hDC, X, Y, lpString, nCount)

This function writes a character string on the specified display, using the currently selected font. The starting position of the string is given by the X and Y parameters.

Parameter	Type/Description
hDC	HDC/Identifies the device context.
X	int/Specifies the logical X-coordinate of the starting point of the string.
Y	int/Specifies the logical Y-coordinate of the starting point of the string.
lpString	LPSTR/Points to the character string that is to be drawn.
nCount	int/Specifies the number of characters in the string.
Return Value	The return value specifies whether or not the string is drawn. It is nonzero if the string is drawn; otherwise, it is zero.
Comments	Character origins are defined to be at the upper left corner of the character cell. By default, the current position is not used or updated by this function. However, an application can call the SetTextAlign function with the wFlags parameter set to TA_UPDATECP to permit *Windows* to use and update the current position each time the application calls TextOut for a given device context. When this flag is set, *Windows* ignores the X and Y parameters on subsequent TextOut calls.

THIRD PARTY DYNAMIC LINK LIBRARIES

In addition to using *Windows* API DLLs, you can access Third Party DLLs written in several languages. You call Third Party DLLs just like you call *Windows* API functions. The C language has dominated *Windows* programming for so long that DLL development is geared toward that language. The procedures and documentation in some Third Party DLLs are documented in C language syntax. To call procedures in Third Party DLLs from *Visual Basic*, you must translate C language syntax declare statement into valid *Visual Basic* syntax declare statements. A list of common C language declarations and their *Visual Basic* equivalents are given in *Visual Basic Programmer's Guide* (Microsoft Press, 1993, pp. 570–571). A mini C primer to convert declarations from C to their *Visual Basic* equivalents is provided by Arnson, Rosen, Waite, and Zuck (1992). Specifics of interfacing C with Fortran programs are provided by Ribar (1993, pp. 123–160). A Third Party DLL written specifically for *Visual Basic* usually supplies a file containing the correct *Visual Basic* declarations for the procedures in the DLL and provides *Visual Basic* code examples.

Muscle, QuickPak Scientific for Windows, and *3D Graphic Tools* are excellent examples of Third Party DLLs. *Muscle* ($199; MicroHelp, Inc., 4359 Shallowford Industrial Parkway, Marietta, GA 30066) is a collection of assembly language routines you can use to speed up and add functionality to your *Visual Basic* application. The DLL contains over 600 routines (display, string, array, memory, mouse, keyboard, file and device, directory, disk, system, error handling, date and time, bit manipulation, and user interface) that can be invoked exactly like *Visual Basic* subprograms and functions.

QuickPak Scientific for Windows ($149; Crescent Software, Inc., 11 Bailey Ave., Ridgefield, CT 06877) adds sophisticated numerical analysis computing capabilities to *Visual Basic*. It contains routines for linear algebra, differential equations, interpolation and curve fitting, fast, Fourier transforms, numerical integration, statistics, and vectors and matrices.

3D Graphic Tools ($149; Microsystem Options, P.O. Box 95167, Seattle, WA 98145–2167) is a complete set of color 3-D drawing and rendering functions packaged as a C language DLL. Function capabilities include contour, spline and arbitrary surface, grid, label, and wireframe drawings. Full viewpoint object orientation and placement control, camera/lens metaphor, parallel and perspective projection, sophisticated lighting, flat and Phong shading, hidden surface removal, translucent objects, picture and textural mapping, shadows, and anti-aliasing are supported. An application starter kit written in *Visual Basic* explains how you can use the C functions from the *Visual Basic* environment.

CUSTOM DYNAMIC LINK LIBRARIES

You cannot create your own DLLs with *Visual Basic*, but you can create custom DLLs that can be called from *Visual Basic* with other programming languages including Microsoft Fortran, version 5.1 ($299); Microsoft *Visual C++,* version 1.0 ($499); WATCOM Fortran 9.1 (719; WATCOM Systems, Inc., 415 Phillip Street, Waterloo, Ontario, Canada); Borland *Pascal with Objects,* version 7.0 ($359); and GFA-Basic for *Windows*, Pro-Version ($395; GFA Software Technologies, Inc., 27 Congress Street, Salem, MA 01970).

GFA-Basic supports direct *Windows* API calls and large arrays. Detailed information about creating and using DLLs with Microsoft Fortran and WATCOM Fortran are provided by Ribar (1993, pp. 95–122).

A Microsoft Fortran version 5.1 DLL can only provide computational services and perform file input and output; it cannot perform screen input and output, nor does it support *Windows* API directly (WATCOM Fortran 9.1 provides access to the complete set of *Windows* API functions). To create a DLL you must compile the Fortran source code with the appropriate compiler options (Aw and Gw must be included), link the resulting object code with the appropriate run-time libraries, and provide the linker with information about routines that are exported from the DLL and imported to the caller (.DEF). A module-definition file (.DEF) is a text file that describes the name, attributes, exports, imports, system requirements, and other characteristics of the DLL. In calling a custom DLL, the libname in the Declare statement can be a file specification with a path, such as C:\WIN\EFD.DLL. A typical FL (compiler) command for creating a DLL is as follows: FL/c/FPi/Gt/Os/Aw/Gw A:EFD.FOR. A typical LINK command for creating a DLL is as follows: LINK EFD.OBJ, EFD.DLL, NUL, /NOD IDLLFEW.LIB, EFD.DEF.

The sample module definition text file code in which MTR is the Fortran routine in EFD.FOR to be called from *Visual Basic* is as follows:

```
;EFD.DEF              Definition file for EFD.DLL
LIBRARY              EFD.DEF
DESCRIPTION          'DLL for EFD program'
EXETYPE              WINDOWS 3.1
PROTMODE
CODE                 PRELOAD MOVEABLE DISCARDABLE
DATA                 PRELOAD MOVEABLE
HEAPSIZE             1024
EXPORTS
MTR
WEP
```

THIRD PARTY CUSTOM CONTROLS AND TOOLS

The Toolbox of the Standard Edition *Visual Basic,* version 3.0 ($199), contains the following 22 tools (called controls) to draw controls on forms: picture box, label, text box, frame, command button, check box, option button, combo box, list box, horizontal and vertical scroll bars, timer, drive list box, directory list box, file list box, shape, line, image, data, grid, OLE, and common dialog (dialog boxes for operations such as opening, saving, and printing files or selecting colors and fonts). A menu control can be accessed from the Menu Design Window. In addition, custom controls are available in the Professional Edition *Visual Basic,* version 3.0, and from several third parties. A custom control is an extension to the Standard Edition *Visual Basic* Toolbox that provides new functionality to an application when it is added to a program. You can write a custom control with a programming language other than *Visual Basic* as explained in the Control Development Guide of the Professional Edition *Visual Basic.*

The Professional Edition *Visual Basic,* version 3.0 ($495), contains the following custom controls in addition to those contained in the Standard Edition: 3-D check box, 3-D command button, 3-D frame, 3-D group push button, 3-D option button, 3-D panel, 3-D group push button, animated button, communications, gauge, graph, key status, MAPI sessions, MAPI messages, masked edit, multimedia MCI, outline, pen BEdit, pen HEdit, pen ink on bitmap, pen-on-screen keyboard, picture clip, and spin button.

The Standard and Professional Editions of *Visual Basic* 3.0 contain the Microsoft *Access* version 1.1 database engine, support for multiple data formats, visual data control and other data-aware controls (text box, check box, label, picture, and image), OLE container control, OLE 2.0 automation support, and Setup Wizard. The Professional Edition of *Visual Basic* 3.0 contains a Microsoft SQL Server, SYBASE SQL Server, Oracle ODBC drivers, ODBC support with scrollable cursor, Programmatic data access layer (database object, table object, etc.), integrated report writer (Crystal Reports version 2.0 for *Visual Basic*), professional data-aware controls (3-DPanel and Checkbox, and Masked Edit), outline control, report control, and data-aware control development kit.

Many other custom controls are offered by a host of third parties. For example, *3-D Widgets* (available from Sheridan Software Systems, 65 Maxess Road, Melville, NY; $88) is a three-library superset of the 3-D custom controls distributed with the Professional Edition of *Visual Basic.* The first library, Widgets/1, provides 3-D versions of the standard option buttons, check boxes, command buttons, frames, panels, and ribbon button. These controls are included in the Professional Edition *Visual Basic* as THREED.VBX. Widgets/2 provides 3-D versions of list boxes, combo boxes, drive list boxes, directory list boxes, and file list boxes. Widgets/3 provides 3-D-appearing menus including menu buttons.

Data Widgets (available from Sheridan Software Systems, 65 Maxess Road, Melville, NY; $129) is a set of data-bound controls (DataGrid, DataDropDown, Enhanced Data Control, DataOption Button, Data Button, and DataCombo) for designing database application front ends. A fully editable bound data grid similar in look and functionality to the Microsoft *Access* data grid can be developed without a line of code.

MicroHelp 3-D Gizmos (available from MicroHelp, Inc., 4359 Shallowford Industrial Parkway, Marietta, GA 30066; $99) includes 3-D label and text box custom controls not provided with *Visual Basic* or *3-D Widgets,* as well as other useful 3-D custom controls.

VB/Magic Controls (available from AJS Publishing, Inc., P.O. Box 83220, Los Angeles, CA 90083; $149) includes roll-ups, sliders, cascading menus, 3-D effects, floating tool palettes, and image-based buttons. An Object Editor constructs reusable buttons, toolbars, palettes, and sliders.

HighEdit (available from MicroHelp, Inc., 4359 Shallowford Industrial Parkway, Marietta, GA 30066; $249) is a WYSIWYG word processor custom control that can be integrated into a *Visual Basic* application. *HighEdit* has the ability to edit multiple documents and to print documents using *TrueType*. It supports multiple fonts, multiple colors, search and replace, justification, text properties, line spacing, and ASCII file import/export. *HighEdit* comes with a ruler, choice of visibility for tabs, and paragraph marks and control characters. There is no text box 64K limitation as there is with *Visual Basic;* instead the actual limit is 2 gigabytes.

Spread/VBX 2.0 (available from FarPoint Technologies, 585A SouthLake Boulevard, Richmond, VA 23236; $245) is a full-featured spreadsheet custom control that can be integrated into a *Visual Basic* application. *Spread/VBX* supports the following features: GUI operations, single or multiple cell operations, data operations, database connectivity, printing, data events, database events, and spreadsheet interface designer. *Spread/VBX* supports the following cell types: data, time, float, integer, formatted PIC, combo-box, bitmaps, icons, or push buttons. Cells may be locked and have their own color and font. Editing is done within the cell. Functions are provided that perform block cell I/O and turn off and on the redrawing of the spreadsheet. The user can specify mathematical functions that support relative/absolute cell referencing. An entire spreadsheet can be written to or read from a file. The user has complete control over what to print as well as how it will appear. *Spread/VBX* supports 32,000 rows and 32,000 columns. The user can resize rows and columns and select a range of cells.

Visual/db (available from AJS Publishing, Inc., P.O. 83220, Los Angeles, CA 90083; $149) is an excellent example of a third party *Visual Basic* form that enables a database manager to be integrated into a *Visual Basic* application. *Visual/db* is a complete relational database management system that creates, reads, and writes industry-standard dBase-type database (DBF), index (NDX), and memo (DBT) files (single user) or reads and writes files for multi-users. It is implemented entirely in *Visual Basic* source code. *Visual/db* sorts records in any order and searches by full/partial key, exact, next, previous, first, or last. With *Visual/db* the user can create a database, display its structure, build a new index, browse, edit, and append records.

Several third party enhanced development environment tools for *Visual Basic* are available. For example, *VBAssist 3.0* (available from Sheridan Software, 65 Maxess Road, Melville, NY; $179) is an add-on that can be used to edit forms, modules, ASCII text, and Code Librarian files with a "visual clipboard", display a menu of available property types and select a type whose properties are to be shown, display a code window to display a menu of code items from a Code Librarian group and select a code item for insertion into the code window, display and change the order of control's TabIndex, and align and size controls automatically. A form/picturebox scroll control lets the user design large forms that can be scrolled in both design and run modes. A data assistant simplifies the task of linking bound controls to database fields using drag and drop. A FormWizard automatically creates a *Visual Basic* form from a database table or query, complete with navigation bar, text boxes, and labels for each field, as well as adding fully functional new, delete, update, and refresh buttons.

TrackDeck (available from Dashboard Software, 4 Louis Ave., Monsey, NY 10952; $129) is an enhanced *Visual Basic* source code debugger add-on. With *TrackDeck* you can show a variable's value in a graphic display, chart memory allocation, plot the variable's changes over time in a graph, and display a variable's statistics (such as rate of growth and maximum value) on the variable's graph.

Test for Windows (available from Microsoft Corporation, P.O. Box 3011, Bothell, WA 98041–3011; $299) automates the testing of *Visual Basic* applications. This noninvasive tool creates and runs automated test scripts that simulate mouse and keyboard input and check for correct results. Unattended tests automatically log results and intercept events such as unrecoverable application errors. APIs and DLLs can be tested.

PinPoint VB (available from Avanti Software, Inc., 3790 El Camino Real, Suite 347, Palo Alto, CA 94306; $99) dramatically reduces debugging time by providing a separate window and real-time, customized view of a program's flow and execution. It sets conditional breakpoints on combinations of variables and functions, including multiple concurrent programs, and captures program flow data to disk for analysis of user activity.

VB Compress (available from Tailored PCs, 20 Cedar Street, Charlestown, MA 02129; $99.95) integrates analysis, Xref, optimization, and debugging; identifies all unreferenced, duplicate, and obsolete code; creates optimixed copy and new EXE without unnecessary bytes; and packages and maintains code in standard libraries.

Visual Solutions (from Borland International, Inc., 100 Borland Way, Scotts Valley, CA; $99.95) is a set of 25 word processing, communications, spreadsheet, database, image editor, 3-D chart, animation button, gadget, slider, clock, alarm, and spinner custom controls.

Visual Tools (available from Evergreen Technologies, Main Street, P.O. Box 795, Castine, ME 04421; $495) contains custom controls and DLLs to display and manipulate complex scientific and engineering data in *Visual Basic* programs.

5   *DESIGNMOD* USER'S MANUAL

INTRODUCTION

DesignMod has a Manager window (hereafter referred to as the Manager), a Map menu, 83 subsidiary dialog box windows (hereafter referred to as dialog boxes), two code modules, and numerous message boxes. Example menus, dialog boxes, and message boxes are presented in Appendix B. The Manager is the hub of activity throughout a program session. The Manager contains a Title Bar with the window title; a Control-Menu box; a Minimize button; a Maximize button; a Menu Bar listing available menus; a drop-down menu space providing information about program version and platform, copyright, and authorship; and a mouse pointer.

The Manager Menu Bar menus are File, Edit, Interpret, View, Draw, and Help. Choosing one of these menu items displays a drop-down menu with commands or options. Choosing a drop-down menu item displays a dialog box or a cascading menu containing options. Selecting a cascading menu item displays a dialog box or a cascading menu containing additional options. Selecting an additional cascading menu option displays a dialog box.

Dialog boxes temporarily overlay the Manager, prompt for data entry and option choice, and contain controls such as text boxes, command buttons, option buttons, picture boxes, spreadsheet boxes, and list boxes. You can use command buttons to store data and option choices in the computer memory; open, read or write, and close a file; make calculations; display data; open another dialog box; or close a dialog box. You can use Option buttons to select a numerical format, array type, display type, coordinate type, or calculation type. You can use picture boxes to view cross sections and plan view maps and array graphics, spreadsheet boxes to view array tables; and list boxes to choose colors, line widths, line styles, fonts, file names, and conversion units. Message boxes contain brief messages about dialog boxes and controls and error traps. The message box remains on the screen until you choose its OK button.

DesignMod does not provide screen, printer, or mouse drivers but relies on Microsoft *Windows* drivers. Screen, printer, mouse, and other device settings are controlled by *Windows*, not *DesignMod*. With the *Windows* Setup program, you install screen fonts to display text on the screen and the printer driver(s) for your printer(s). The printer driver provides access to printer fonts and instructions for your printer. You also choose a port to establish a software connection between your computer and the printer, and you choose the active printer and settings that *DesignMod* will use with the installed printer driver. You also install the mouse driver and set the mouse port. *DesignMod* uses the printer default font settings. You can change your system's configuration by using *Windows* Control Panel or Setup.

You can change the default printer by using *Windows* Control Panel. Switch from *DesignMod* to the *Windows* Program Manager by pressing "Ctrl + Esc" to display the Task List. Click the Program Manager option or use the arrow keys to move the mouse pointer to the Program Manager option. Click the Switch To button or tab to the Switch To button and press "Enter". After switching to the Program Manager, open the Main Group and click the Control Panel icon or use the arrow keys to move the mouse pointer to the Control Panel icon and press "Enter".

Click the Control Panel Settings menu and the Printers option or Press "Alt + S" and "Alt + P" to display a Printers dialog box. In the Installed Printers box, select the printer you want to use as the default printer by clicking the desired option or using the arrow keys to move the mouse pointer to the desired option and pressing "Enter". Select the SetUp button to change printer settings by clicking the SetUp button or tab to the SetUp button and press "Enter". A dialog box appears prompting for information about printer changes. After choosing the printer and changing any printer settings, click or tab to the Close button to return to the Control Panel menu. Close the Control Panel by clicking the Settings menu and the "Exit" option, or press "Alt + S" and "Alt + X" and return to the Program Manager. Press "Alt Esc" to return to *DesignMod*.

DesignMod does not have its own file manager, printer manager, control panel, or accessories. You can use the *Windows* File Manager to open, move, copy, delete, or rename a file; create a directory; and format or copy a disk. You can use the *Windows* Print Manager to install and configure a printer, change printer settings (paper source, paper size, orientation, and resolution), choose the default printer, look at and change the print queue, and pause and resume printing. You can use the *Windows* Control Panel to customize the mouse operation, install and configure printers and fonts, and control the print quality. You can use the *Windows Paintbrush* accessory to elaborate and modify if necessary and print cross section or plan view images created with *DesignMod*. For example, you can use the *Paintbrush* Paint Roller to color unit or layer areas between well logs. You can use the *Windows* Calculator accessory to do selected scientific and statistical calculations.

DesignMod has the state-of-the-art three-dimensional gray NeXT look which is best when the color of the *Windows* screen elements Active Title Bar and Highlight menu is gray and the color of the Active Bar text is black. If necessary, you can change the color of these elements by choosing the *Windows* Control Panel Color settings (see "Changing a Color Scheme" section of Chapter 5 in the Microsoft *Windows User's Guide*).

WINDOWS PRIMER

Before setting up *DesignMod,* you should become familiar with the mouse and *Windows* if you have not already done so. You should be able to rapidly operate *Windows* Program Manager and the various parts of a *Windows* application, such as menus, dialog boxes, list boxes, option buttons, command buttons, scroll bars, and the Control Panel menu box. Become familiar with the mouse and *Windows* by reading the chapter "Basic Skills" in the *Windows User's Guide* and practicing. A brief *DesignMod*-oriented *Windows* primer is provided herein for your convenience.

You use the mouse to point to parts of the *Windows* Program Manager or the *DesignMod* Manager, menus, dialog boxes, picture boxes, spreadsheet boxes, and message boxes; set the focus on text boxes that will receive the next data entry; choose menus; select options; choose buttons; and select items in list boxes. These functions are performed by clicking (quickly pressing and releasing the left mouse button once) or double-clicking (quickly pressing and releasing the left mouse button twice in succession).

You can shrink the Program Manager to an application icon at the lower left of the screen to reduce screen clutter by pressing "Alt + Esc" to open the Program Manager, pressing and releasing "Alt", pressing and releasing the spacebar to open the Program Manager Control Panel menu box, using arrow keys to highlight Minimize and pressing "Enter". Press "Alt + Esc" to return to *DesignMod* Manager. You can restore the Program Manager to its original size by double-clicking the Program Manager icon or by pressing and releasing "Alt" and pressing and releasing the spacebar to open the Program Manager Control menu, using arrow keys to highlight Restore and pressing "Enter".

If you accidently hide a window in use, press "Alt + Esc" or "Alt + F6" to restore the window. You can quit Program Manager and *Windows* and return to the DOS command by clicking the File menu, clicking the Exit Window File menu option, and clicking the "OK" button in the confirmation dialog box, or by pressing "Alt + F", pressing "X", pressing "Enter", and clicking the "OK" button in the confirmation dialog box.

You can use the Program Manager Run command to start the *DesignMod* Setup program or to run *DesignMod.* Choose the File menu in Program Manager by clicking "File" or by pressing "Alt + F". A drop-down menu appears. Choose Run by clicking "Run" or by pressing "R". A Run dialog box appears requesting the name of the command file. Type the path and name of the program, such as A:SETUP or C:\DM\DM, in the Command Line text box. Click the "OK" button or press "Enter" to close the dialog box and run the program.

You can run *DesignMod* from its group icon. Choose the *DesignMod* group icon within the borders of the Program Manager by double-clicking the icon or by pressing "CTRL + Tab" to highlight the icon and pressing "Enter". A Group window appears containing the *DesignMod* icon. Choose that icon to run *DesignMod* by double-clicking the icon or by pressing "Enter". You can also run *DesignMod* by clicking the *Windows* Program Manager menu or pressing "Alt + W" to open the *Windows* drop-down menu. Click the "More" drop-down menu item or press "M". Click the *DesignMod* cascading menu item or use the arrow keys to move the mouse pointer to *DesignMod* and press "Enter".

DesignMod Manager has a Minimize button and a Maximize button located at the far right of the Title Bar. Pointing to and clicking the Minimize button — a small triangle pointing down — shrinks the Manager to an icon at the lower left of the screen. The Manager can be returned to its original size by double-clicking the *DesignMod* icon or by pressing and holding down "Alt", pressing "Tab" until the *DesignMod* icon is highlighted, and releasing "Alt". The Manager can also be shrunk to an icon by clicking the *DesignMod* Manager Control-Menu box located at the far left of the Title Bar and clicking the "Minimize" drop-down menu item.

Pointing to and clicking the Maximize button — a small triangle pointing up — fills the screen with the Manager if it does not already do so. The Manager can be returned to its original size using the Control-Menu box. The Control-Menu box, located to the far left of the Title Bar at the top of the screen, is a small square with a narrow rectangle inside of it. Open the Control-Menu box by pointing to the Control-Menu box and clicking the mouse or by pressing "Alt + Spacebar". The Control-Menu box has the following options: Restore, Move, Minimize, Maximize, Close, and Switch To. Select (highlight) options in the box by clicking a name, by pressing arrow keys to select a name and pressing "Enter", or by typing the appropriate underlined letter and pressing "Enter". A dimmed Control-Menu box option means that command is not available (i.e., is disabled) at this time. The key combination after certain Control-Menu box command names is a shortcut for choosing the command. Choosing the Move command in the Manager Control-Menu box causes the mouse pointer to change to a four-sided arrow which can be used with the arrow keys to move the *DesignMod* Manager. Press "Enter" when movement is complete. Choosing the "Close" command exits *DesignMod.* Choosing the "Switch To" command results in the appearance of the Task List dialog box that *Windows* maintains to enable you to switch between applications running on the *Windows* desktop and rearrange windows and icons on the desktop. Double-clicking on the Control-Menu Box will exist *DesignMod.*

A dimmed item in the *DesignMod* Menu Bar indicates that another menu must be chosen before the dimmed menu is available. A drop-down menu appears after you choose a menu by clicking the menu or by pressing "Alt" +

the underlined letter in the desired menu. You can select an option on the drop-down menu by clicking the option, pressing the underlined letter in the desired menu item's name, or pressing arrow keys until the desired menu is highlighted and then pressing "Enter". You can cancel a drop-down menu by clicking the menu or pressing "Esc".

You can select a cascading option by clicking the option or pressing "Alt" + the underlined letter in the desired menu item's name or by using the arrow keys to highlight the desired menu item and then pressing "Enter". An ellipsis (…) after a menu item indicates that a dialog box will appear when the option is selected. A triangle at the right side of the menu item name leads to a cascading menu which lists additional options

DesignMod dialog boxes have no Maximize or Minimize button and their sizes cannot be changed. The Control-Menu box, located to the far left of the dialog box Title Bar at the top of the screen, is a small square with a narrow rectangle inside of it. The Control-Menu box can be used with or without a mouse to move and close the dialog box and to switch to the Task List. You open the Control-Menu box by pointing to the Control-Menu box and clicking the mouse or by pressing "Alt + Spacebar". Select (highlight) commands in the box by clicking a name, by pressing arrow keys to select a name and pressing "Enter", or by typing the appropriate underlined letter and pressing "Enter". A dimmed Control-Menu box command means that command is not available (i.e., is disabled) at this time. The key combination after certain Control-Menu box command names is a shortcut for choosing the command.

Dialog boxes prompt for data entry and contain one or more controls (text boxes, list boxes, picture boxes, spreadsheet boxes, option buttons, and command buttons) and are temporary windows that have input, output, calculation, option selection, or other command functions. After data entry is complete and you have validated input, you choose a command button to carry out an action, cancel the action, choose additional options, or exit from the dialog box. Dimmed buttons are unavailable until another button is chosen.

You can navigate around a dialog box by moving the mouse pointer to a control and clicking the control or by pressing "Tab" to move forward through controls or "Shift + Tab" to move in the opposite direction. Direction keys can be used to move from one option to another. You select a button by clicking the button or by pressing "Tab" to move to the desired button and then pressing the spacebar or "Enter". If you accidently hide a dialog box in use, press "Alt + F6" to restore the window. Some dialog boxes must be closed before action can occur in another dialog box or a menu. If you attempt to activate another form or a menu, the computer will beep and no further action will occur until you close the dialog box.

You enter (type) information in blank white rectangular text boxes. You give a text box the focus by clicking the mouse anywhere inside the text box or by pressing "Tab". An insertion point (flashing vertical bar) appears at the far left of the box when you move the mouse pointer into the text box. You can type two kinds of information directly into text boxes: numerals (values) and characters (text). Values can be formatted in standard decimal or exponential form. You can change text box entries by using standard editing keys, including arrow, backspace, Ins, and Del. Text is typed starting at the insertion point. Existing text can be erased by using the Delete or Backspace key. Complete the entry (store text in the computer's memory) by clicking on an "Enter" or "OK" button or by moving to an Enter or OK button with Tab and then pressing "Enter".

You can end a dialog before completion and close the dialog box by clicking the "Cancel" button, pressing the "Esc" key, double-clicking a dialog box Control-Menu box, clicking the Control-Menu box and then clicking the "Close" option, or pressing "Alt + Spacebar" to open the Control-Menu box and pressing "C" to choose close.

A list box displays a column of available choices. If more choices are available than will fit into the list box, a scroll bar is provided so that you can rapidly move up and down in the list box. Scroll bars are used to navigate within the list box window. Select a single item from a list box by clicking the scroll arrows until the choice appears in the list box, then double click the item desired. You can also select a single item by pressing "Tab" to move to the list box (watch for the dashed frame to appear inside the top item in the list), using arrow keys to scroll to the desired item, and pressing "Enter" to choose the item.

Option buttons appear in dialog boxes as lists of mutually exclusive items. You can choose only one item from the list. The selected option button contains a white dot, and unavailable options are dimmed. Select an option button by clicking the desired button or by pressing "Tab" to move to the group of options; use arrow keys to move to the desired button and press "Enter".

DesignMod displays data arrays in a spreadsheet box (ledger sheet) with column and row cells and horizontal and vertical scroll bars. Editing of the spreadsheet is not supported. Selecting (highlighting) one or more cells and placing the data for the cell(s) onto the Clipboard for export to another application program is supported. You can use the scroll bars to move through and view a data array that is too large to be displayed all at once in the grid box. You can scroll through one row or column at a time by clicking the scroll bar arrow, clicking in the scroll bar on either side of the scroll box, dragging the scroll box, or using the arrow keys.

Before you can place data for cell(s) onto the Clipboard, you need to select one or more cells. When you select a single cell, a border appears around it. When you select a range of cells (set of adjacent cells), the grid box extends the highlight by changing the background color of the selected cells. You can select a range of cells by clicking on the cell at the upper left corner of the range and then dragging the mouse to the opposite corner of the range. Press and hold down the mouse button while moving the mouse pointer over the appropriate cell(s). Clicking the column heading extends the highlight over an entire column. Clicking the row heading extends the highlight over an entire row.

You can also select a range of cells by using arrow keys to move the mouse pointer to the upper left corner of the range; press and hold down "Shift" and use arrow keys to move the mouse pointer to the opposite corner of the range. Deselect a range of cells by clicking in any cell or pressing the spacebar key. You can deselect a range of cells and set the focus in the upper left corner cell by pressing "Home". After selecting one or more adjacent cells, you can place the data from the cell(s) onto the Clipboard by choosing the Clip command on the spreadsheet dialog box.

You can view, save, retrieve, and delete the contents of the Clipboard by using the *Windows* Clipboard Viewer. Press "Ctrl + Esc" to display the *Windows* Program Manager Task List. Click the Program Manager option from the list of running applications or use arrow keys to move the mouse pointer to the Program Manager option. Tab to the Switch To button and press "Enter" or click on the "Switch To" button. The Program Manager appears. To open the Main Group, press "Alt + W" or click the *Windows* menu, press "M" or click the More drop-down menu option, click "Main" in the Select Window list box or use arrow keys to move the mouse pointer to Main in the Select Window list box, click the "OK" button or tab to the OK button, and press "Enter". In the Main Group, click on the Clipboard Viewer icon or use arrow keys to move the mouse pointer to the Clipboard Viewer icon and press "Enter". The contents of the Clipboard appear.

You can save the contents of the Clipboard to a file using the Clipboard Viewer File menu by clicking the "Save As" option or by pressing "Alt + A". In the Save File dialog box that appears, use the Backspace key to clear the filename text box, type a valid filename with an extension .CLP in the filename text box, and click the "OK" button or tab to the OK button and press "Enter". You can retrieve a Clipboard file with an extension .CLP using the Clipboard Viewer File menu by clicking the "Open" button or pressing "Alt + O". In the Open File dialog box that appears, select the file you want to retrieve by clicking the desired option and the "OK" button or by using arrow keys to move the mouse printer to the desired option, tab to the OK button, and press "Enter". You can clear the contents of the Clipboard using the Clipboard Viewer Edit menu by clicking the "Delete" option or by pressing "Alt + D". Exit the Clipboard Viewer by clicking the File menu and clicking the "Exit" option, or tab to the File menu and press "Alt + X". Press "Ctrl + Esc" to return to the *DesignMod* Manager and the spreadsheet grid box. The *Windows* Program Manager Task List appears. Click the *DesignMod* option from the list of running applications or use arrow keys to move the mouse pointer to the *DesignMod* option. Tab to the Switch To button and press "Enter", or click on the "Switch To" button.

SETUP

Before setting up *DesignMod,* check your hardware to see that it meets the following *DesignMod* basic requirements: a MS-DOS system computer with a 286 microprocessor or greater, a math coprocessor, a hard drive with at least 3 megabytes free, a 1.44-MB 3½-inch floppy disk driver, 4 MB of RAM or greater, an enhanced graphics adapter (EGA) monitor or better, a Microsoft or compatible mouse, MS-DOS 3.0 or newer, and *Windows* 3.0 or newer.

Start *Windows* if it is not already running and close any open applications. Place the *DesignMod* disk in drive A or drive B. From the Program Manager File menu choose "Run". In the Command Line box type "A:Setup" (or "B:Setup"). Choose the "OK" button and follow instructions on the screen. The Setup program installs *DesignMod* files in appropriate directories. If you do not desire to use the automatic Setup program and wish to manually install *DesignMod* files, follow the instructions provided below.

If *Windows* is not already running, start it by typing WIN at the MS-DOS prompt (C:\) and press "Enter". The Program Manager window appears on the screen. Insert the *DesignMod* disk in drive A. Start the File Manager by choosing the File Manager icon in the Main group (double-click the icon or move the selection cursor to the File Manager icon and press "Enter"). Select the A drive in the directory window by clicking the A drive icon or by pressing Tab to move to the drive icons, using the arrow keys to select the A drive icon, and pressing the Spacebar. The contents of the *DesignMod* disk appear in the directory window. Check the directory window to make sure that the contents include the following files (but not necessarily in this order): DM.EXE, VBRUN200.DLL, GRID.VBX, THREED.VBX, SS3D2.VBX, SS3D3.VBX, DM1C.DAT, DM1G.DAT, DM1H.DAT, DM1M.DAT, DM2G.DAT, DM2H.DAT, DM2S.DAT, DM3G.DAT, DM4GK.DAT, DM4GT.DAT, DM4S.DAT, DM6G.DAT, DM7A.DAT, DM7L.DAT, DM7S.DAT, DM9.TXT, SETUPKIT.DLL, VER.DL_, SETUP1.EXE, SETUP.EXE, and SETUP.LST. If one or more of these files are not on the disk, notify the publisher.

Create a new directory named C:\DM using the *Windows* File Manager. From the File menu, choose "Create Directory". A Create Directory dialog box appears. Type the full path of the new directory (C:\DM) in the Name text box and press the "OK" button. Select all of the files through DM9.TXT in the drive A contents list in the directory window by clicking the first file and pressing and holding down "Shift" while you click the last file, or press "Tab" to move the selection cursor to the drive A contents list, use the arrow keys to move to the first file, and press and hold down "Shift" while using the arrow keys to select the remaining files. Copy the selected files to the new directory C:\DM by choosing "Copy" from the File menu. A Copy dialog box appears. Type C:\DM in the "To" text box and click the "Yes To All" button. All of the selected files on the *DesignMod* disk are now copied to the new directory

C:\DM. Select the C drive in the directory window by clicking the C drive icon or by pressing "Tab" to move to the drive icons, using the arrow keys to select the C drive icon, and pressing the Spacebar. Select the DM directory icon. The contents of the DM directory appear in the directory window.

Remove the *DesignMod* disk from drive A and insert a high density (1.44-MB) *DesignMod* backup disk in drive A. From the File Manager Disk menu, choose "Format Disk". A Format Disk dialog box appears. Select the letter "A" in the "Disk In" box, select the size of the disk you want to format in the Capacity box, and choose the "OK" button. A confirmation dialog box appears. Choose the "Yes" button. After the *DesignMod* backup disk is formatted, a dialog box appears. Choose the "No" button.

Select all of the files in the directory C:\DM contents list in the directory window by clicking the first file and pressing and holding down "Shift" while you click the last file or by pressing "Tab" to move the selection cursor to the directory C:\DM contents list, using the arrow keys to move to the first file, and pressing and holding down "Shift" while using the arrow keys to select the remaining files. Copy the selected files to the formatted disk in drive A by choosing "Copy" from the File menu. A Copy dialog box appears. Type A: in the "To" text box and choose the "Yes To All" button. All of the files on the *DesignMod* diskette are now copied to the *DesignMod* backup disk.

Quit the File Manager by choosing "Exit" from the File menu or by pressing "Alt + Spacebar" to open the Control-Menu box; choose "Close".

You can add a *DesignMod* group to the Program Manager, thereby facilitating the running of *DesignMod.* From the Program Manager File menu, choose "New". A New Program Object dialog box appears. Select the "Program Group" option and choose the "OK" button. A Program Group Properties dialog box appears. In the Description text box, type DESIGNMOD and choose the "OK" button. The opened *DesignMod* Group window appears. Create a program item for *DesignMod* inside the *DesignMod* Group window by choosing "New" in the File menu. A New Program Object dialog box appears. Select the "Program Item" option and choose the "OK" button. A Program Item Properties dialog box appears. In the Description text box, type DESIGNMOD. In the Command Line text box, type C:\DM\DM (the name of the program file DM.EXE). Choose the "OK" button, and the new program item appears inside the DesignMod Group window.

You can change the size of the *DesignMod* Group window by pointing to a border or corner (the mouse pointer changes to a two-headed arrow). Press and hold the left mouse button down and drag the border or corner until the group window is the size you want. After sizing the group window, you can move the group window to a desired position by dragging the group window. You can shrink the *DesignMod* Group window to a group icon by clicking the "Minimize" button (the box with a down arrow to the right of the Title Bar). You can move the Group icon to a new position by moving the mouse cursor to the icon, pressing and holding down the left mouse button, and dragging the icon. To save new Program Manager screen settings without quitting Program Manager, press and hold down the "Shift" key and then choose "Exit Windows" from the File menu.

If you run into trouble setting up *DesignMod*, read appropriate sections of the *Windows User's Guide*. You can copy *DesignMod* files to directories other than C:\DM. If you do, be aware that the *DesignMod* file DM.EXE, *Visual Basic* run-time file VBRUN200.DLL, and the Custom Control files GRID.VBX, THREED.VBX, SS3D2.VBX, and SS3D3.VBX should be copied to the same directory. *DesignMod* will not run without proper access to the run-time and Custom Control files. You can format a disk, create the C:\DM directory, and copy files using the MS-DOS commands FORMAT, MD, and COPY. Read the MS-DOS *User's Guide* for instructions on how to use these commands.

You may wish to edit your AUTOEXEC.BAT and/or CONFIG.SYS file(s). In this case, choose "Run" from the File menu in *Windows* Program Manager. Type SYSEDIT in the text box and press "Enter". Click the window in System Editor that contains the file to be edited. Edit the file using the text editing techniques described in the *Windows* Primer. Save the file and choose "Exit" from the File menu to close the System Editor. Restart *Windows* (reboot your system) so that the changes that you made can take effect.

The author has taken due care in developing the *DesignMod* disk, including research and testing to ensure its accuracy and effectiveness. Neither the author nor the publisher makes any warranty of any kind, expressed or implied, with regard to the performance of the *DesignMod* diskette and associated User's Manual. No warranties, expressed or implied, are made by the author or publisher that the *DesignMod* program is free of error, or is consistent with any particular standard of merchantability, or that it meets the reader's requirements for any particular application. The *DesignMod* disk and the User's Manual are used at the user's own risk, relying solely upon his/her own inspection of the *DesignMod* disk and User's Manual contents and without reliance upon any representation or description concerning the *DesignMod* disk.

In no event shall the author or the publisher be liable for incidental or consequential direct or indirect damages, including lost profits, lost savings, and interruption of business, in connection with or arising from the use, inability to use, furnishing, or performance of the *DesignMod* disk and User's Manual, even if the author or the publisher has been advised of the possibility of damages, or any claim by any other party.

Support for the *DesignMod* disk is the User's Manual and is not available via telephone or mail. The author and the publisher do not offer consulting services to assist users in the solution of groundwater modeling problems using *DesignMod*.

A license to use the enclosed software and documentation is granted to the original purchaser of this book. Copies may be made for backup purposes only. Copies made for any other purpose are expressly prohibited, and adherence to this requirement is the sole responsibility of the purchaser. Installation of this product on more than one computer, if there is the chance that both copies will be used simultaneously, is prohibited. This restriction also extends to the installation on a network server if more than one workstation will be accessing the product.

START AND EXIT

You should read the entire User's Manual before starting *DesignMod*. For those of you who just can't wait and want to try out *DesignMod* before reading the entire User's Manual, read the Online Help section of this manual. After starting *DesignMod*, choose the Help menu in the Manager Menu Bar and select and read each of the cascading help options. Follow the steps outlined in Exercise DM3 to see how *DesignMod* works.

DesignMod is represented by a group icon along the lower edge of the Program Manager. Start *DesignMod* from *Windows* by choosing the group icon; a Program Manager *DesignMod* Group window appears. Choose the *DesignMod* program-item icon in the group window. The DesignMod Manager window appears and you can use *DesignMod*. Shrink the Program Manager to a group icon using its Minimize button to reduce screen clutter.

You can also start *DesignMod* by choosing the *Windows* Program Manager File menu and the Run option. When the Run dialog box appears, type the appropriate specification, such as C:\DM\DM.EXE, in the Command Line text box and choose "OK".

DesignMod is closed by selecting the Manager file menu and then choosing the "Exit" command in the cascading file menu or by pressing "ALT + F" and then pressing "ALT + X". Upon exiting *DesignMod*, a message appears on the screen. You can choose the "OK" button to complete the exit and return to *Windows* Program Manager or choose the "Cancel" button to return to the *DesignMod* Manager to save a file.

ONLINE HELP

DesignMod supports an extensive online Help system. The Manager Menu Bar and each cascading menu contain a Help option, and each dialog box contains a Help button. Choosing the Manager Menu Bar "Help" menu displays a cascading menu with the following options: About Help, Menus, Coordinates, Arrays, Error Traps, and Keyboard. Selecting "About Help" displays a message box with a brief description of the Help system. Selecting "Menus" displays a message box with information about menus. Selecting "Coordinates" displays a message box with information about site and model systems. Selecting "Arrays" displays a message box with information about the types of arrays supported and array dimension limitations. Selecting "Error Traps" displays a message box with information about error trapping, reporting, and correction. Selecting "Keyboard" displays a message box with information about using keys to recover hidden windows.

Choosing a cascading menu Help item displays a message box with information common to each cascading menu option. Choosing a dialog box Help button displays a message box with information about the function of the dialog box buttons and other controls and array and block dimension limitations. Help Message boxes partially cover the Manager or a dialog box. After reading the message, choose the Help Message box "OK" button to close (hide) the Help Message box and uncover the dialog box.

OPERATING LIMITS

DesignMod has a number of operating limits which if exceeded will result in an error message and require input modification before further program operation. Many limits are associated with array integer dimensions. Operating limits are listed in Table 1.

If the contents of a gridded triplet array are to be viewed in a tabular spreadsheet and dialog box, the number of columns multiplied by the number of rows must not exceed 1600.

ERROR TRAPS

Three types of run-time errors are trapped by *DesignMod:* file handling, invalid information entry, and calculation errors. When a run-time error occurs, an Error Message box appears describing the error and halting execution, sometimes returning a predefined error number and information about handling the error. In the case of invalid information entry, choose the Error Message box "OK" button to clear appropriate text boxes and reset the focus so that information can be retyped. In the case of improper file handling or calculation error, you will be requested to return to the Manager to correct the situation and try again. In any event, error traps prevent most unintentional unloading (ending abruptly) of *DesignMod* with data loss. Most run-time errors occur because of illogical information entry or conceptual modeling mistakes. Every attempt has been made to trap errors known to the author, but it is impossible to trap all conceivable user mistakes.

The most common error code numbers reported in Error Message boxes and their explanations are listed in Table 2.

TROUBLESHOOTING

DesignMod runs on the *Windows* platform, which is a working environment surrounding MS-DOS, the operating system that controls basic functions within your computer. Therefore, any trouble you have setting up or using *DesignMod* can be associated with MS-DOS, *Windows*, *DesignMod*, or a

Table 1. Operating Limits

Operating Parameter	Limit
Block numbers	1–100
Browse files	ASCII; 32,000 bytes
File catalog	ASCII; 32,000 bytes
Fill styles	8
Code name characters	3
Colors	12 (contour)
	14 (features)
Column numbers	1–200
Constituent numbers	1–25
Contour numbers	1–12
Data element numbers	1–500
Description characters	24
Doublet files	ASCII
Font names	3
Font sizes	6
Graph label characters	20
Graph log cycles	10
Grid files	ASCII or binary
Line styles	7
Log data	X- and Y-axis > 0
Mean element numbers	2–500
Name characters	24
Number of time steps	1–100
Number of wells along cross section	1–25
Point numbers	1–500
	5–500 (interpret)
	1–100 (information)
Range numbers	1–12 (zonal diagram)
Row numbers	1–200
Semilog data	X-axis > 0
Sink or source numbers	2–500
Stress period numbers	1–100
Triplet files	ASCII
Triplet origin	Upper left in memory, lower left files
Unit characters	1–12
Unit numbers	25
	10 (concentration)
Vertice numbers	4–100
Site base map coordinates	Positive
Well log coordinates	Positive
Well numbers	100
	25 (cross section)
Zones	14

Table 2. Error Code Numbers

Error Code Number	Explanation	Error Code Number	Explanation
4	Out of data	58	File already exists
5	Illegal function call	59	Bad record length
6	Overflow	61	Disk full
7	Out of memory	62	Input past end of file
9	Subscript out of range	63	Bad record number
11	Division by zero	64	Bad file name
14	Out of string space	65	File previously loaded
24	Device timeout	67	Too many files
25	Device fault	68	Device unavailable
27	Out of paper	70	Permission denied
28	Out of stack space	71	Disk not ready
40	Variable required	72	Disk-media error
50	Field overflow	75	Path/file access error
51	Internal error	76	Path not found
52	Bad file name or number	460	Invalid clipboard format
53	File not found	482	Printer error
54	Bad file mode	520	Can't empty clipboard
55	File already open	521	Can't open clipboard
56	Field statement active	30009	Too many columns and
57	Device I/O error		rows

combination thereof. It is remotely possible that *DesignMod* will fail to load or run properly and stop responding to your system, and/or your system may lock up after displaying an error message. When this occurs, carefully read and record the error message and make a note of what you were doing when the trouble occurred. Follow the error message instructions for exiting the program. You should not to try to continue *DesignMod* after an error occurs; if you do, part of the memory may be corrupted.

Try to solve your problem (which may be invalid data entry) by carefully reading the entire *DesignMod* User's Manual, Chapter 4 (Troubleshooting MS-DOS) in the *Getting Started Manual* for MS-DOS, and Chapter 4 (Troubleshooting) in the *Getting Started with Microsoft Windows Manual*. If the trouble persists, you should call your hardware dealer and/or Microsoft Product Support Services (described in MS-DOS and *Windows* manuals) for help.

FILE MENU

Choosing the *DesignMod* File menu displays a drop-down menu with six options: Catalog, for creating, modifying, and viewing lists of file names and contents; Browse, for viewing array file contents and formats; Load, for importing a file; Save, for exporting a file; Exit, for closing *DesignMod;* and Help, for viewing information about the other five options. Choosing one of these options displays a word processor window, dialog box, message box, or a cascading menu with additional options. Choosing one of the cascading menu options displays a dialog box or another cascading menu. Choosing the "Help" option displays a message box with information about the file menu. Choosing the message box "OK" button closes the message box and returns to the Manager. Choosing the "Cancel" button in any dialog box associated with the File menu ends the dialog before completion and returns to the Manager. The specifications (filenames) for files to be browsed, loaded, or saved must have extensions. *DesignMod* does not automatically back up files. Positive site base map coordinates are supported.

Catalog

Keeping track of a large collection of files is no small task. Catalog enables you to organize and maintain a list of file names and associated brief content descriptions, thereby assisting you in managing these files. Catalog is a customized multiline text editor for viewing and editing a list of files and can greatly reduce the time you spend in finding any particular file. ASCII files with less than 32,000 bytes are supported. If you attempt to load a file with more than 32,000 bytes, a warning message appears informing you that your file is too large to edit.

Selecting "Catalog" displays a File Catalog word processor window. The File Catalog window contains a Title Bar with the window title, a Menu Bar listing available menus, a Text Box Heading Bar with two labels (File Name and Content Description), a Text Box with a flashing mouse cursor, and a Control-Menu Bar. The File Catalog Menu Bar menus are File, Edit, and Help. Choosing one of these menus displays a drop-down menu with options or displays a help message.

Selecting the "File" menu displays a drop-down menu with four options: New, Load, Save, and Exit. Selecting "New" erases the current content of the text box so that a new file catalog can be created. Selecting "Exit" closes the File Catalog window and returns to the Manager.

Selecting "Load" or "Save" displays a Specify Filename dialog box prompting for the name of the file to be loaded for viewing and editing or saved. You can examine your file system and select among system files with the Drive List box, Directory List box, Wildcard Filename Extension box, and File List Box. Alternatively, you can specify a valid name for the catalog file to be loaded or saved in one of two ways: type the appropriate filename in the Filename text box and choose the "OK" button to load or save the catalog file with the typed filename; or use the Drive List box scroll bar arrow to select the desired drive, select the desired directory in the Directory List box, edit the default *.* extension in the Wildcard Filename Extension box and choose the Extension button (thereby limiting your Filename List Box search), use the Filename List box scroll bar arrow to select the desired filename, and choose the "OK" button to load or save the catalog file.

When you load a catalog file or start a new catalog, the text insertion point (flashing bar) appears in the upper left corner of the text box. The text that you type starts filling the text box at the insertion point. You can move the insertion point by moving the mouse cursor to a desired place and clicking the mouse button once or by using the arrow keys. You can correct typing mistakes by using Backspace to back up the insertion point and undoing the mistake. You can skip one or more spaces by pressing the spacebar. You can end a line by pressing "Enter" or leave a blank line by pressing "Enter" twice. Create columns of text by pressing "Tab", thereby using preset tabs to align text to the left.

You can move around the catalog by using the text box scroll bars or arrow keys. Click in the scroll bars to view parts of the text box, or drag the scroll box to move to more than one screenful at a time. Press "PgDn" or "PgUp" to move one screenful at a time. Press "Ctrl + End" or "Ctrl + Home" to move directly to the end or beginning of the catalog. Press "Home" to move directly to the beginning of a line and "End" to move directly to the end of a line. Use the arrow keys to move left or right one character.

Selecting the Edit menu displays a drop-down menu with three options: Cut, Copy, and Paste. Cut and Copy are enabled if text is selected (high-lighted), and Paste is enabled if no text is selected (highlighted). You can edit (insert or delete) selected text by choosing either "Cut", thereby deleting the selected text from the text box and placing it onto the Clipboard, or "Copy", making a copy of the selected text and placing it onto the Clipboard. Place the insertion point where you want the text that was previously placed on the Clipboard to appear and choose Paste, which pastes text from the Clipboard into the text box. Rearrange items in the catalog by cutting and pasting the items. Information that you cut or copy onto the Clipboard remains there until you clear the Clipboard, cut or copy another piece of information, or exit File Catalog.

Before your edit text, you need to select (highlight) it. You can select text by pressing and holding down the mouse button while moving the mouse pointer over the appropriate text or by pressing "Shift" and the left or right arrow keys. The background color of the selected text is changed. You can cancel the selection mode by clicking the mouse button or by pressing an arrow key. You can select a line by moving the insertion mouse cursor to the left margin of the text box and double-clicking the mouse button or by pressing "Shift" and "End".

Selecting "Help" displays a Help Message dialog box. After reading the message, choose "OK" to close the dialog box and return to the File Catalog window. Use *Windows* WRITE and a file to obtain a hard copy of a catalog. Exit the File Catalog and return to the Manager by choosing the File menu "Exit" option.

Browse

Choosing "Browse" displays a Specify Filename dialog box prompting for the name of the file to be browsed. The file must have an ASCII format. You can examine your file system and select among available files with the Drive List box, Directory List box, Wildcard Filename Extension box, and File List box. Alternatively, you can specify a valid name for the file to be loaded in one of two ways: (1) type the appropriate filename in the Filename text box and choose the "OK" button to store the filename; or (2) use the Drive List box scroll bar arrow to select the desired drive, select the desired directory in the Directory List box, edit the default *.* extension in the Wildcard Filename Extension box and choose the "Extension" button (thereby limiting your Filename List box search), use the Filename List box scroll bar arrow to select the desired filename, and choose the "OK" button to store the filename.

Choosing the "OK" button displays a Browse dialog box containing a text box with scroll bars. The name of the file appears above the text box. The contents of the file appear in tabular form within the text box. Only the first 32,000 bytes of large files can be viewed using the scroll bars.

Choosing the "Size" button displays a File Fact message box showing the size of the file in bytes. Choosing the "Values" button displays a File Fact message box showing the number of values in the array. This information can be particularly useful if the number of array columns and rows is unknown. You can ascertain the number of array columns and rows in part by inspecting the array file's contents, looking for patterns of repetition of X, Y, or Z values. Choosing the File Fact message box "OK" button closes the dialog box. Choosing the Browse dialog box "OK" button closes the dialog box and returns to the Manager.

Load

Choosing "Load" displays a cascading menu with six options: Log, Grid Spacing, Doublet, Scattered Triplet, Gridded Triplet, and Grid. Log refers to an array of well unit or model layer data, Grid Spacing refers to an array of model grid column or row grid spacings, Doublet refers to an array of XY values, Scattered Triplet refers to an array of scattered point XYZ values with either a lower left or upper left origin, Gridded Triplet refers to an array of grid node XYZ values with either a lower left or upper left origin, and Grid refers to an array of grid node Z values with a model upper left origin.

Selecting "Log" displays a cascading menu with four options: Geology, Hydrogeology, Geochemistry, and Model. Geology refers to well log data based on geologic units, Hydrogeology refers to well log data based on hydrogeologic units, Geochemistry refers to well log data based on hydrogeologic units and selected constituents, and Model refers to model layer data. Well locations are defined by site base map coordinates with a lower left origin.

Selecting "Scattered Triplet" or "Gridded Triplet" displays a cascading menu with two options: Upper Origin and Lower Origin. Upper Origin refers to an array with a model upper left origin, and Lower Origin refers to a site base map array with a lower left origin. Selecting the "Grid" option displays a cascading menu with two options: ASCII and Binary. ASCII refers to an array ASCII-formatted file, and Binary refers to a binary-formatted file.

Selecting "Load" and any of the cascading menu items with "…" at the end displays a Specify Filename dialog box prompting for the name of the file in which well or model data are stored. You can examine your file system and select among available files with the Drive List box, Directory List box, Wildcard Filename Extension box, and File List box. Alternatively, you can specify a valid name for the file to be loaded in one of two ways: (1) type the appropriate filename in the Filename text box and choose the "OK" button to store the filename; or (2) use the Drive List box scroll bar arrow to select the desired drive, select the desired directory in the Directory List box, edit the default *.* extension in the Wildcard File Extension box and choose the Extension button (thereby limiting your Filename List box search), use the Filename List box scroll bar arrow to select the desired filename, and choose the "OK" button to store the filename. Choosing the "OK" button also closes the dialog box and displays a particular Specify File Dimensions data entry dialog box depending on menu selections.

Selecting "Load", "Log", and "Geology"; "Load", "Log", and "Hydrogeology"; or "Load", "Log", and "Model" displays a Specify Filename dialog box which leads to a Specify File Dimensions data entry dialog box prompting for information about the wells or models whose geologic or hydrogeologic data are stored in the file to be imported. Type the number of wells or models, the well or model number, and the number of units or layers whose data are stored in the file to be imported in appropriate text boxes. Choose the "Enter Well Or Model" button to store these entries.

Type the number of the first unit or layer associated with the previously entered well or model number whose data are stored in the file to be imported in the appropriate text box. Choose the "Enter Unit Or Layer" button to store this entry. Choose the "Next Unit Or Layer" button to clear the text box and reset the focus so that numbers for additional units or layers can be entered. This process makes it possible for nonconsecutive numbers to be entered. Entries must correspond to those used to store data in the file to be imported. Choose the "Next Well Or Model" button to clear text boxes and reset the focus so that additional well or model numbers can be entered. Well and model numbers 1–100 and unit or layer numbers 1–25 are supported. Choosing the "OK" button opens, reads, and closes the file; closes the dialog box; and returns to the Manager.

Selecting "Load", "Log", and "Geochemistry" displays a Specify Filename dialog box and leads to a Specify File Dimensions data entry dialog box prompting for information about the wells or models whose geochemical data are stored in the file to be imported. Type the number of wells, the well number, the number of hydrogeologic units, and the number of constituents whose data are stored in the file to be imported in appropriate text boxes. Choose the "Enter Well" button to store these entries. Type the number of the first hydrogeologic unit associated with the previously entered well number whose data are stored in the file to be imported in the appropriate text box. Choose the "Enter Unit" button to store this entry. Choose the "Next Unit" button to clear text boxes and reset the focus so that additional hydrogeologic units numbers can be entered. Type in the appropriate text box the number of the first constituent associated with the previously entered well and unit numbers for which data are stored in the file to be imported. Choose the "Enter Constituent" button to store this entry. Choose the "Next Constituent" button to clear the constituent text box and reset the focus so that numbers for additional constituents can be entered. Your entries must correspond to those used to store data in the file to be imported. Choose the "Next Well" button to clear text boxes and reset the focus so that data for additional wells can be entered. Well and model numbers 1–100, unit or layer numbers 1–10, and constituent numbers 1–25 are supported. Choosing the "OK" button opens, reads, and closes the file; closes the dialog box; and returns to the Manager.

Selecting "Load" and "Grid Spacing" displays a Specify Filename dialog box and leads to a Specify File Dimensions data entry dialog box prompting for information about the model grid spacings stored in the file to be imported. Select the file type (column grid spacings or row grid spacings) by choosing the appropriate option button. Type the left and right column numbers or the upper and lower row numbers for the grid block whose spacings are stored in the file to be imported in appropriate text boxes. Entries must correspond to those used to store data in the file to be imported. Column and row numbers 1–200 are supported. Choosing the "Enter" button stores entries. Choosing the "OK" button opens, reads, and closes the file; closes the dialog box; and returns to the Manager.

Selecting "Load" and "Doublet" displays a Specify Filename dialog box and leads to a Specify File Dimensions data entry dialog box prompting for information about points for which data are stored in the file to be imported. Type the first and last point numbers in appropriate text boxes. Entries must correspond to those used to store data in the file to be imported. Point numbers 1–500 are supported. Choosing the "Enter" button stores entries. Choosing the "OK" button opens, reads, and closes the file; closes the dialog box; and returns to the Manager.

Selecting "Load", "Scattered Triplet", and "Upper" displays a Specify Filename dialog box and leads to a Specify File Dimensions data entry dialog box prompting for information about points for which data are stored in the file to be imported. Type the first and last point numbers in appropriate text boxes. Entries must correspond to those used to store data in the file to be imported. Point numbers 1–500 are supported. Choosing the "Enter" button stores entries.

Choosing the "OK" button opens, reads, and closes the files; closes the dialog box; and returns to the Manager.

Selecting "Load", "Scattered Triplet", and "Lower" displays a Specify Filename dialog box and leads to a Specify File Dimensions data entry dialog box prompting for information about points for which data are stored in the file to be imported. Type the first and last point numbers and the reference frame upper and lower Y-coordinates in the appropriate text boxes. Reference frame refers to Y-coordinates of upper and lower boundary lines of the array grid with a lower left origin. Entries must correspond to those used to store data in the file to be imported. Point numbers 1–500 are supported. Choosing the "Enter" button stores entries. Choosing the "OK" button opens, reads, and closes the file; transforms the grid origin from lower left to upper left; closes the dialog box; and returns to the Manager.

Selecting "Load", "Gridded Triplet", and "Upper or Lower" displays a Specify Filename dialog box and leads to a Specify File Dimensions data entry dialog box prompting for information about the grid block for which data are stored in the file to be imported. Type the upper left and lower right corner column and row numbers in appropriate text boxes. Entries must correspond to those used to store data in the file to be imported. Column and row numbers 1–200 are supported. Choosing the "Enter" button stores entries. Choosing the "OK" button opens, reads, and closes the file; transforms the grid origin from lower left to upper left; closes the dialog box; and returns to the Manager.

Selecting "Load", "Grid", and "ASCII or Binary" displays a Specify Filename dialog box and leads to a Specify File Dimensions data entry dialog box prompting for information about the grid block for which data are stored in the file to be imported. Only array grid files with model upper left origins are supported. Type the upper left and lower right corner column and row numbers in appropriate text boxes. Entries must correspond to those used to store data in the file to be imported. Column and row numbers 1–200 are supported. Choosing the "Enter" button stores entries. Choosing the "OK" button opens, reads, and closes the file; closes the dialog box; and returns to the Manager.

Save

All or any portion of any array stored internally can be saved by choosing the File menu option "Save", which displays a cascading menu with six options: Log, Grid Spacing, Doublet, Scattered Triplet, Gridded Triplet, and Grid. Log refers to an array of well unit or model layer data, Grid Spacing refers to an array of model grid column or row grid spacings, Doublet refers to an array of XY values, Scattered Triplet refers to an array of scattered point XYZ values, Gridded Triplet refers to an array of grid node XYZ values, and Grid refers to an array of grid node Z values.

Selecting "Log" displays a cascading menu with four options: Geology, Hydrogeology, Geochemistry, and Model. Geology refers to well log data based on geologic units, Hydrogeology refers to well log data based on hydrogeologic units, Geochemistry refers to well log data based on hydrogeologic units and selected constituents, and Model refers to model layer data.

Selecting "Scattered Triplet" or "Gridded Triplet" displays a cascading menu with two options: Upper Origin and Lower Origin. Upper Origin refers to an array with a model upper left origin, and Lower Origin refers to a site base map array with a lower left origin. Selecting the "Grid" option displays a cascading menu with three options: CSV, LSSV, and Binary. CSV refers to a comma-separated value ASCII-formatted array file, LSSV refers to a listed space-separated value ASCII-formatted array file with six or less values on a line, and Binary refers to a binary-formatted file. Selecting "LSSV" displays a cascading menu with two options: Real and Integer. Real refers to grid array values with decimal points, and Integer refers to grid array values without decimal points.

Selecting "Save", "Log", and "Geology"; "Save", "Log", and "Hydrogeology"; or "Save", "Log", and "Model" displays a Specify File Dimensions data entry dialog box prompting for information about the wells or models whose geologic or hydrogeologic data are to be stored in a file. Type the number of wells or models, the well or model numbers, and the number of well units or model layers in appropriate text boxes. Type a valid name such as C:\DM\DM1G.DAT for the file in which data are to be stored in the Save As text box. Choose the "Enter Well Or Model" button to store entries. In the appropriate text box, type the number of the first unit or model layer associated with the previously entered well or model number whose data are to be stored in the file. Choose the "Enter Unit Or Layer" button to store this entry. Choose the "Next Unit Or Layer" button to clear the text box and reset the focus so that data for additional units or layers can be entered. Entries must be consistent with any array dimensions stored internally by using the Edit menu or by importing a file. Choose the "Next Well Or Model" button to clear text boxes and reset the focus so that additional well or model numbers can be entered. Well and model numbers 1–100 and unit or layer numbers 1–25 are supported. Choosing the "OK" button opens, writes to, and closes the file; closes the dialog box; and returns to the Manager.

Selecting "Save", "Log", and "Geochemistry" displays a Specify File Dimensions data entry dialog box prompting for information about the wells whose geochemical data are to be stored in a file. Type the number of wells, the well number, the number of hydrogeologic units, and the number of constituents whose data are to be stored in a file in appropriate text boxes. Type a valid name such as C:\DM\DM1C.DAT for the file in which data are to be stored in the Save As text box. Choose the "Enter Well" button to store entries.

Type the number of the first hydrogeologic unit associated with the previously entered well number whose data are to be stored in a file in the appropriate text box. Choose the "Enter Unit" button to store this entry. Choose the "Next Unit" button to clear the unit text box and reset the focus so that data for additional units can be entered. In the appropriate text box, type the number of the first constituent associated with the previously entered well and unit numbers for which data are to be stored in a file. Choose the "Enter Constituent" button to store this entry. Choose the "Next Constituent" button to clear the constituent text box and reset the focus so that data for additional constituents can be entered. Choose the "Next Well" button to clear text boxes and reset the focus so that data for additional wells can be entered. Entries must be consistent with array dimensions stored internally by using the Edit menu or by importing a file. Well numbers 1–100, unit numbers 1–10, and constituent numbers 1–25 are supported. Choosing the "OK" button opens, writes to, and closes the file; closes the dialog box; and returns to the Manager.

Selecting "Save" and "Grid Spacing" displays a Specify File Dimensions data entry dialog box prompting for information about the model grid spacings to be stored in a file. Select the file type (column grid spacings or row grid spacings) by choosing the appropriate option button. In the appropriate text boxes, type the left and right column numbers or the upper and lower row numbers for the grid block whose spacings are to be stored in a file. Entries must be consistent with array dimensions stored by using the Edit menu or by importing a file. Column and row numbers 1–200 are supported. Select the format type (CSV or LSSV) by choosing the appropriate option button. Type a valid name such as C:\DM\DM1GS.DAT for the file in which data are to be stored in the Save As text box. Choosing the "Enter" button stores entries. Choosing the "OK" button opens, writes to, and closes the file; closes the dialog box; and returns to the Manager.

Selecting "Save" and "Doublet" displays a Specify File Dimensions data entry dialog box prompting for information about points for which data are to be stored in a file. Type the first and last point numbers in appropriate text boxes. Entries must be consistent with array dimensions stored by using the Edit menu or by importing a file. Point numbers 1–500 are supported. Type a valid name such as C:\DM\DM1D.DAT for the file in which the data are to be stored in the Save As text box. Choosing the "Enter" button stores entries. Choosing the "OK" button opens, writes to, and closes the file; closes the dialog box; and returns to the Manager.

Selecting "Save", "Scattered Triplet", and "Upper" displays a Specify File Dimensions data entry dialog box prompting for information about points for which data are to be stored in a file. Type the first and last point numbers in appropriate text boxes. Entries must be consistent with array dimensions stored internally by using the Edit menu or by importing a file. Point numbers 1–500 are supported. Type a valid name such as C:\DM\DM1STU.DAT for the file in which the data are to be stored in the Save As text box. Choosing the "Enter" button stores entries. Choosing the "OK" button opens, writes to, and closes the file; closes the dialog box; returns to the Manager.

Selecting "Save", "Scattered Triplet", and "Lower Origin" displays a Specify File Dimensions data entry dialog box prompting for information about points for which data are to be stored in a file. Type the first and last point numbers in appropriate text boxes. Entries must be consistent with array dimensions stored internally by using the Edit menu or by importing a file. Point numbers 1–500 are supported. Type a valid name such as C:\DM\DM1STL.DAT for the file in which the data are to be stored in the Save As text box. Choosing the "Enter" button stores entries. Choosing the "OK" button opens, writes to, and closes the file; transforms the grid origin from lower left to upper left; closes the dialog box; and returns to the Manager.

Selecting "Save", "Gridded Triplet", and "Upper Origin" displays a Specify File Dimensions data entry dialog box prompting for information about the grid block for which data are to be stored in a file. Type the upper left and lower right corner column and row numbers in appropriate text boxes. Entries must be consistent with array dimensions stored internally by using the Edit menu or by importing a file. Column and row numbers 1–200 are supported. Type a valid name such as C:\DM\DM1GTU.DAT for the file in which the data are to be stored in the Save As text box. Choosing the "Enter" button stores entries. Choosing the "OK" button opens, writes to, and closes the file; closes the dialog box; and returns to the Manager.

Selecting "Save", "Gridded Triplet", and "Lower Origin" displays a Specify File Dimensions data entry dialog box prompting for information about the grid block for which data are to be stored in a file. Type the first and last point numbers and the reference frame upper and lower Y-coordinates in the appropriate text boxes. Reference frame refers to upper and lower boundary lines of the grid with a lower left origin. Entries must be consistent with array dimensions stored internally by using the Edit menu or by importing a file. Column and row numbers 1–200 are supported. Type a valid name such as C:\DM\DM1GTL.DAT for the file in which the data are to be stored in the Save As text box. Choosing the "Enter" button stores entries. Choosing the "OK" button opens, writes to, and closes the file; closes the dialog box; and returns to the Manager.

Selecting "Save", "Grid", and "CSV"; "Save", "Grid", "LSSV", and "Real"; "Save", "Grid", LSSV", and "Integer"; or "Save", "Grid", and "Binary" displays a Specify File Dimensions data entry dialog box prompting for information about the grid block for which data are to be stored in a file. Type the upper left and lower right corner column and row numbers in appropriate text boxes. Entries must be consistent with array dimensions stored internally by using the Edit menu or by importing a file. Column and row numbers 1–200 are supported. Type a valid name such as C:\DM\DM1GR.DAT for the file in which data are to be stored in the Save As text box. Choosing the "Enter" button stores entries. Choosing the "OK" button opens, writes to, and closes the file; closes the dialog box; and returns to the Manager.

EDIT MENU

The Edit menu can be used to convert individual data array values and entire data arrays from one system of units to another; develop or edit a model geologic, hydrogeologic, or geochemical well log or layer database; create model grid spacing arrays with uniform or variable grid spacings; create a new data array with uniform or variable values; and edit existing data arrays by adding new values to the array, modifying selected values of the array, and combining several arrays into a larger array. Existing data array files must be loaded using the File menu before they can be edited. All or any portion of any data array created, developed, or edited with the Edit menu can be saved using the File menu. Positive site base map coordinates are supported.

The geologic well log database contains the well number, well X- and Y-coordinates, unit numbers, unit names, unit codes, and unit top and bottom elevations. The hydrogeologic well log database contains the well number, well X- and Y-coordinates, unit numbers, unit names, unit codes, unit top and bottom elevations, and unit water level elevations. The geochemical well log database contains the well number, unit numbers, constituent numbers, constituent names, constituent codes, and constituent concentrations. The model log contains the model number, layer numbers, layer names, layer codes, and layer top and bottom elevations. Wells or models are identified internally not only with your well or model numbers, but also with numbers 1 through the number of wells or models in the order that you enter well or model data using the Edit menu. A relationship is established between numbers 1 through the number of wells or models and your original well or model numbers. Your original well or model numbers are stored in arrays that are saved.

The zonal block overlay method of editing Z value data arrays is supported, wherein the array grid is divided into a number of blocks (discrete zones), each having a different uniform Z value. Block numbers 1–100 are supported. Blocks comprise one or more cell nodes of a finite-difference model grid and are rectangles. A block overlay can cover one cell node, one column or part of a column of cell nodes, one row or part of a row of cell nodes, several cell columns and rows, or the entire grid. A single value is specified for each block overlay so that cell nodes within a block share a common value. Block overlays can overlap so that some values previously specified for certain cell nodes are replaced.

Blocks are defined by specifying the column and row numbers of upper left and lower right block corners. Existing Z values within a block overlay can be replaced or modified by adding to, subtracting from, multiplying by, or dividing existing values by a specified value. Replacements or modifications can be applied to cell nodes whose values are within a Z value criteria range. A search is performed using the following criteria range: cell node Z value is greater than specified lower range value and equal to or less than the specified upper range value. Z values that meet these conditions are identified and edited.

Selecting the "Edit" menu displays a drop-down menu with the following options: Units, Log, Grid Spacing, Scattered Triplet, Z Or XYZ, Doublet, and Help. Unit refers to unit conversion, Log refers to a well or model log database, Grid Spacing refers to a column or row grid spacing array, Scattered Triplet refers to an array of XYZ values for scattered control points, Z or XYZ refers to a gridded array of Z values or a gridded array of X- and Y-coordinates and Z values, and Doublet refers to an array of XY values. Choosing the Cancel button in any Edit dialog box ends the dialog before completion and returns to a dialog box or the Manager.

Units

Selecting "Units" displays a cascading menu with three options: English To Metric, Metric To English, and Other. English To Metric refers to the conversion of a value from an English unit to the corresponding metric unit. Metric To English refers to the conversion of a value from a metric unit to an English unit. Other refers to the conversion of a value from one English unit to another English unit or from one metric unit to another metric unit. Selecting any one of these options displays a Specify Units data entry dialog box prompting for information about the conversion dimension, type of unit to be used in value conversion, and value to be converted.

Choose the desired dimension in the Dimension List box using the vertical scroll bar arrows to search the list of available dimensions. When the desired dimension is highlighted in the Dimension List box, choose the "Dim" button to select that dimension. The Units List box is filled after the desired dimension is selected. View unit abbreviations by using the scroll bar arrows in the Abbreviations text box. Choose the desired units in the Units List box using the vertical scroll bar arrows to search the list of available unit conversions. When the desired unit conversion is highlighted in the Units List box, choose the Units button to select that unit conversion.

The conversion factor (value or equation) appears on the screen after the desired unit conversion is selected. Type the value to be converted in the Value text box. Choosing the "Convert" button converts the value and displays the converted value. Make additional conversions using the selected conversion factor by choosing the "Next" button, thereby clearing text boxes and resetting the focus.

You must preload or pre-edit a grid, gridded triplet (Triplet-G), scattered triplet (Triplet-S), or doublet data array before choosing the Array button. Choosing the Array button prior to preloading or pre-editing an array will yield unexpected results. Choosing the "Array" button displays a Specify Conversion

Dimensions data entry dialog box prompting for information about the array type and predetermined conversion factors. Select the type of the preloaded or pre-edited array. If "Grid" was selected, type the grid array's upper left and lower right corner column and row numbers in appropriate text boxes. Type the appropriate conversion factor in the Z Value Conversion text box. If "Triplet-G" was selected, type the triplet array's upper left and lower right corner column and row numbers in the appropriate text boxes. Type the X-coordinate, Y-coordinate, and Z value conversion factors in the appropriate text boxes. If "Triplet-S" was selected, type the scattered triplet array's first and last point numbers in appropriate text boxes. Type the X-coordinate, Y-coordinate, and Z value conversion factors in appropriate text boxes. If Doublet was selected, type the doublet array's first and last point numbers in appropriate text boxes. Type the X- and Y-coordinate conversion factors in appropriate text boxes. Choosing the "OK" button converts each value in the data array, closes the dialog box, and returns to the Specify Units dialog box. Choosing the "OK" button in the Specify Units dialog box closes that dialog box and returns to the Manager.

Log

Selecting "Log" displays a cascading menu with four options: Geology, Hydrogeology, Geochemistry, and Model. Geology refers to well log data based on geologic units, Hydrogeology refers to well log data based on hydrogeologic units, Geochemistry refers to well log data based on hydrogeologic units and selected constituents, and Model refers to model layer data.

Selecting "Log" and "Geology" displays a Specify Geology Dimensions data entry dialog box prompting for information about the well for which geologic data are available. Type the well number, number of geologic units, X- and Y-coordinates, and X- and Y-coordinate units in appropriate text boxes. Choose the "Enter Well" button to store these entries. Well numbers 1–100 and 12 coordinate unit characters are supported. Type the unit number, unit name, unit code name, unit top elevation, unit base elevation, and elevation unit associated with the previously entered well number in the appropriate text boxes. Unit numbers 1–25, 24 unit name characters, 3 unit name code characters, and 12 elevation unit characters are supported. Choosing the "Enter Unit" button stores these entries. Choosing the "Next Unit" button clears Unit text boxes and resets the focus so that data for additional units can be entered. Choosing the "Next Well" button clears Well and Unit text boxes and resets the focus so that data for additional wells can be entered. Choosing the "OK" button closes the dialog box and returns to the Manager.

Selecting "Log" and "Hydrogeology" displays a Specify Hydrogeology Dimensions data entry dialog box prompting for information about the well for which hydrogeologic data are available. Type the well number, number of hydrogeologic units, X- and Y-coordinate units, and X- and Y-coordinates in the appropriate text boxes. Choose the "Enter Well" button to store these entries. Well numbers 1–100 and 12 coordinate unit characters are supported. Type the unit number, unit name, unit code name, unit top elevation, unit base elevation, unit water level elevation, and elevation unit associated with the previously entered well number in the appropriate text boxes. If appropriate, type 999.99 in the unit water level elevation text box to signify no available data. Unit numbers 1–25, 24 unit name characters, 3 unit name code characters, and 12 elevation unit characters are supported. Choosing the "Enter Unit" button stores these entries. Choosing the "Next Unit" button clears Unit text boxes and resets the focus so that data for additional units can be entered. Choosing the "Next Well" button clears Well and Unit text boxes and resets the focus so that data for additional wells can be entered. Choosing the "OK" button closes the dialog box and returns to the Manager.

Selecting "Log" and "Geochemistry" displays a Specify Geochemical Dimensions data entry dialog box prompting for information about the well for which geochemical data are available. Type the well and unit numbers in the appropriate text boxes. Well numbers 1–100 and unit numbers 1–10 are supported. The unit numbers must correspond to hydrogeologic unit numbers. Choose the "Enter" button to store these entries. Type the constituent number, constituent name, constituent code name, constituent concentration, and constituent concentration unit associated with the previously entered well and unit numbers in the appropriate text boxes. Constituent numbers 1–25, 24 unit name characters, 3 unit name code characters, and 12 constituent concentration unit characters are supported. Choosing the "Enter Constituent" button stores these entries. Choosing the "Next Constituent" button clears text boxes and resets the focus so that data for additional constituents can be entered. Choosing the "Next Unit" button clears text boxes and resets the focus so that data for additional units can be entered. Choosing the "Next Well" button clears text boxes and resets the focus so that data for additional wells can be entered. Choosing the "OK" button closes the dialog box and returns to the Manager.

Selecting "Log" and "Model" displays a Specify Model Dimensions data entry dialog box prompting for information about the model for which data are available. Type the model number and number of layers in the appropriate text boxes. Choose the "Enter Model" button to store these entries internally. Model numbers 1–100 and layer numbers 1–25 are supported. Type the layer number, layer name, layer code name, layer top elevation, layer base elevation, layer water level elevation or 999.99, and elevation unit associated with the previously entered model number in the appropriate text boxes. Layer numbers 1–25, 24 layer name characters, 3 layer name code characters, and 12 elevation unit characters are supported. Choosing the "Enter Layer" button stores these entries. Choosing the "Next Layer" button clears Layer text boxes and resets the focus so that data for additional layers can be entered. Choosing the "Next Model" button clears Model and Layer text boxes and resets the focus so that data for additional models can be entered. Choosing the "OK" button closes the dialog box and returns to the Manager.

Grid Spacing

DesignMod supports uniform and variable model grid spacings and the zonal block method of entering grid spacings. The model grid is divided into one or more column blocks each having the same column grid spacing and one or more row blocks each having the same row grid spacing. Data for at least one column block and one row block must be entered. The number of grid columns along the X-axis, the number of grid rows along the Y-axis, and the number of column and row blocks must be determined before grid spacings can be properly specified.

Choosing "Grid Spacing" displays a Specify Grid Spacing Dimensions data entry dialog box requesting information about grid column and row grid spacings. The column block number 1 appears at the top of the dialog box. Type column block number 1 first and last column numbers and the column grid spacing in appropriate Block Column Data text boxes. Choosing the "Enter Col" button stores column and grid spacing entries. Choosing the "Next Col" button clear text boxes, resets the focus, and increments the block number so that the grid spacing for another column block can be entered. Type column block 2 first and last column numbers and the column grid spacing in the appropriate text boxes. Choosing the "Enter Col" button stores these entries. Choosing the "Next Col" button clears text boxes, resets the focus, and increments the block number so that the grid spacing for another column block can be entered. Repeat this process until all column block grid spacings have been entered.

Choosing the "Row" button changes the block number to 1, indicating that data for row block 1 can be entered, and resets the focus in the Block Row Data text box. Type row block number 1 first and last row numbers and the row grid spacing in the appropriate Block Row Data text boxes. Choosing the "Enter Row" button stores these entries. Choosing the "Next Row" button clears text boxes, resets the focus, and increments the block number so that the grid spacing for another row block can be entered. Type row block 2 first and last row numbers and the row grid spacing in the appropriate text boxes. Choosing the "Enter Row" button stores these entries. Choosing the "Next Row" button clears text boxes, resets the focus, and increments the block number so that the grid spacing for another row block can be entered. Repeat this process until all row block grid spacings have been entered. Type the model grid upper left node X- and Y-coordinates and left column, right column, upper row, and lower row numbers in the appropriate text boxes at the bottom of the dialog box. Choosing the "Enter Grid" button stores model grid data. Choosing the "OK" button closes the dialog box and returns to the Manager.

Scattered Triplet

Selecting "Scattered Triplet" displays a Specify Surface Scattered Point Dimensions data entry dialog box prompting for information about the scattered points and displaying the point number. Type the X- and Y-coordinates of the point and the Z value for the point in the appropriate text boxes. Point numbers 4–500 are supported. Choosing the "Enter" button stores these entries. Choosing the "Next" button clears text boxes and resets the focus so that data for additional points can be entered. Choosing the "OK" button closes the dialog box and returns to the Manager.

Z or XYZ

Selecting "Z Or XYZ Array" displays a cascading menu containing six edit operator options: New, Add, Subtract, Multiply, Divide, and Search. Selecting any one of these options except Search displays a Specify Edit Block Dimensions data entry dialog box prompting for information about the size of the edit block and displaying the block number. Edit block numbers 1–100 are supported. Type the upper left and lower right edit block corner column and row numbers in appropriate text boxes. Column and row numbers 1–200 are supported. Identify the type of array to be edited by selecting one of two Array Type option buttons (Gridded Triplet or Grid). If "Gridded Triplet" was selected, specify whether or not the grid spacing should be created or edited by selecting one of two Grid Spacing option buttons (Edit or Do Not Edit). If "New" was selected, type the Z value for the block in the Value Or Operator Factor text box. If the "Add", "Subtract", "Multiply", or "Divide" menu item was selected, type the operator factor in the Value Or Operator Factor text box. Choosing the "Enter" button stores these entries. Choosing the "Next" button clears text boxes, increments the block number, and resets the focus so that data for additional blocks can be entered. If the "Do Not Edit Grid Spacing" option button was selected, choosing the "OK" button closes the dialog box and returns to the Manager. If the "Edit Grid Spacing" option button was selected, choosing the "OK" button displays a Specify Grid Spacing Dimensions data entry dialog box prompting for information about grid column and row grid spacings (see Grid Spacing section).

Selecting the "Search" option displays a cascading menu containing five edit operator options: Replace, Add, Subtract, Multiply, and Divide. Selecting any one of these options displays a Specify Edit Block Dimensions data entry dialog box prompting for information about the size of the edit block and displaying the block number. Edit block numbers 1–100 are supported. Type the upper left and lower right edit block corner column and row numbers in the appropriate text boxes. Column and row numbers 1–200 are supported. Identify the type of array to be edited by selecting one of two Array Type option buttons (Gridded Triplet or Grid). If "Gridded Triplet" was selected, specify whether or not the grid spacing should be created or edited by selecting one of two Grid Spacing option buttons (Edit or Do Not Edit). If "New" was selected, type the Z value in the Value Or Operator Factor text box. If the "Add", "Subtract", "Multiply", or "Divide" menu item was selected, type the operator factor in the Value Or Operator Factor text box. Type the upper and lower values of the search range in appropriate text boxes. Choosing the "Enter" button stores these entries. Choosing the "Next" button clears text boxes, increments the block number, and resets the focus so that data for additional edit blocks can be entered. If the "Do Not Edit Grid Spacing" option button was selected, choosing the "OK" button closes the dialog box and returns to the Manager. If the "Edit Grid Spacing" option button was selected, choosing the "OK" button displays a Specify Grid Spacing Dimensions data entry dialog box prompting for information about grid column and row grid spacings (see Grid Spacing section).

Doublet

Selecting "Doublet" displays a Specify Point Dimensions data entry dialog box prompting for information about the points whose data are to be edited. Point numbers 1–500 are supported. Type the point number and X- and Y-coordinates and the first and last point numbers of the array in the appropriate text boxes. Choosing the "Enter" button stores these entries. Choosing the "Next" button clears text box boxes and resets the focus so that data for additional points can be entered. Choosing the "OK" button closes the dialog box and returns to the Manager.

INTERPRET MENU

Selecting "Interpret" displays a drop-down menu with the following options: Area, Mean, Sinks, Surface, Time Step, and Help. Area refers to the calculation of the area enclosed within polygon vertices (see Poole et al., 1981, pp. 53–54). Mean refers to the calculation of the mean value of a control data set (see Walton, 1992, pp. 31–32). Sinks refers to the calculation of the equivalent coordinates and discharge rate of a single sink based on known coordinates and discharge rates of several individual sinks (see Walton, 1992, pp. 30–31). The equivalent coordinates and recharge rate of a single source, based on known coordinates and recharge rates of several individual sources, can also be calculated.

Surface refers to the transformation of XYZ values for scattered control points into an interpolation grid (model cell node) XYZ array or the transformation of a model cell node gridded XYZ array with large grid spacings into a model cell node gridded XYZ array with smaller grid spacings. Triangulation and kriging techniques for interpolating between the scattered control data points are supported. To use the triangulation and kriging techniques, you must first preload or pre-edit a scattered or gridded XYZ triplet array. The interpolation grid overlying the control data points can be uniform or variable. Be aware that calculation time will be long if the interpolation grid contains a large number of columns and rows. In the triangulation technique, interpolation between grid nodes and control data points is based on the inverse distance squared weighted average method involving the closest six control data points (see Walton, 1991, pp. 282–283). The influence of a data point decreases with the distance from the grid point. Calculated Z values can never exceed the Z value range of the data points. The triangulation technique works best when there is no presumed regional trend to the data.

The punctual kriging technique is supported. This technique involves a linear semivariogram model with control data points below a sill (flat region) chosen to represent the semivariogram (see Walton, 1991, pp. 284–288). The best-fit line through control data points is assumed to go through the semivariogram origin. The best-fit line is found with the regression least squares method. Interpolation grid Z values and standard errors of estimate are estimated as a weighted average of the known Z values for the three control data points closest to each grid node. With kriging there can be unexpected results in map areas unsupported by data, and Z values can go beyond the Z value range of the control data points. Kriging works best when there is a presumed regional trend to the data.

More sophisticated kriging software is available from

- U.S. Environmental Protection Agency (EPA), Robert S. Kerr Environmental Research Laboratory, Technical Support Center, P.O. Box 1198, Ada, OK 74821–1198 (GEOPACK)
- U.S. EPA EMSL-LV (EAD), P.O. Box 93478, Las Vegas, NV 89193–3478 (GEO-EAS)
- International Ground Water Modeling Center, Institute for Ground-Water Research and Education, Colorado School of Mines, Golden, CO 80401–1887 (GEO-EAS $50 and GEOPACK $50)
- Scientific Software Group, P.O. Box 23041, Washington, D.C. 20026–3041 (GEOKRIG)

Time Step refers to the calculation of model simulation time steps and critical time values (see Anderson and Woessner, 1992, pp. 204–207, 327). Time discretization together with the selection of grid spacings based on flow and transport critical time increment values are important aspects of model design. Time steps are usually increased as a geometric progression of ratio 1.2 to 1.5 The flow model initial time step (flow critical time increment) should be small enough, or the first few time steps should be ignored, to allow for an explicit formulation of governing finite-difference equations. The transport model time step (transport critical time increment) should be small enough to minimize numerical dispersion errors.

Positive site base map coordinates are supported. Choosing the "Cancel" button in any Interpret dialog box ends the dialog before completion and returns to the Manager.

Area

Choosing "Area" displays a Specify Vertice Dimensions data entry dialog box prompting for information about the area vertices and displaying the vertice number. Type the number of vertices in the appropriate text box. Vertice numbers 1–100 are supported. The number of vertices must be four or more. Type the X- and Y-coordinates for the displayed vertice number in appropriate text boxes. Choosing the "Enter" button stores these entries. Choosing the "Next" button increments the vertice number, clears text boxes, and resets the focus so that data for additional vertices can be entered. Choosing the "Area" button calculates and displays the area enclosed by the vertices. Choosing the "OK" button closes the dialog box and returns to the Manager.

Mean

Selecting "Mean" displays a cascading menu with five options: Arithmetic, Geometric, Harmonic, Vertical, and Horizontal. Arithmetic, geometric, and harmonic mean refer to the calculation of a particular type of mean value for a set of element values. Vertical refers to the calculation of the equivalent vertical hydraulic conductivity of a single layer representing several individual model layers. Horizontal refers to the calculation of the equivalent horizontal hydraulic conductivity of a single zone representing several individual zones of a model layer. Choosing any one of these options displays a Specify Mean Element Dimensions data entry dialog box prompting for information about the elements and displaying the element number. Element numbers 1–500 are supported.

Type the number of elements in the appropriate text box. The number of elements must be at least two. If "Arithmetic", "Geometric", or "Harmonic" was chosen, type the value for the displayed element number in the appropriate text box. Choose the "Enter" button to store this entry. Choosing the "Next" button increments the element number, clears the value text box, and resets the focus so that values for additional elements can be entered. Choosing the "Mean" button calculates and displays the mean of the element values.

If "Vertical" was chosen, type values for the vertical hydraulic conductivity and thickness of the displayed element in appropriate text boxes. Choose the "Enter" button to store these entries. Choosing the "Next" button increments the element number, clears the vertical hydraulic conductivity and thickness text boxes, and resets the focus so that values for additional elements can be entered. Choosing the "Mean" button calculates and displays the equivalent hydraulic conductivity.

If "Horizontal" was chosen, type values for the horizontal hydraulic conductivity and width of the displayed element in appropriate text boxes. Choose the "Enter" button to store these entries. Choosing the "Next" button increments the element number, clears the horizontal hydraulic conductivity and width text boxes, and resets the focus so that values for additional elements can be entered. Choosing the "Mean" button calculates and displays the equivalent horizontal hydraulic conductivity. Choosing the "OK" button closes the dialog box and returns to the Manager.

Sinks

Selecting "Sinks" displays a Specify Sink Or Source Dimensions data entry dialog box prompting for information about the sinks or sources and displaying the individual sink or source number. Sink or source numbers 1–500 are supported. Type the number of sinks or sources in the appropriate text box. The number of sinks or sources must be at least two. Type the individual sink or source X- and Y-coordinates and the discharge or recharge rates for the displayed sink or source number in the appropriate text boxes. Choosing the "Enter" button stores these entries. Choosing the "Next" button increments the individual sink or source number, clears text boxes, and resets the focus so that values for additional individual sinks or sources can be entered. Choosing the "Single" button calculates and displays the X- and Y-coordinates and the discharge or recharge rate of the equivalent single sink or source. Choosing the "OK" button closes the dialog box and returns to the Manager.

Surface

Selecting "Surface" displays a cascading menu with two options: Triangulation and Kriging. Selecting "Triangulation" or "Kriging" displays a Specify Grid Spacing Dimensions data entry dialog box prompting for information about the dimensions of the interpolation grid (model cell nodes) that will overlay the scattered control data points. The column block number 1 appears at the top of the dialog box. Type column block number 1 first and last column numbers and the column grid spacing in the appropriate Block Column Data text boxes. Choosing the "Enter Col" button stores column entries. Choosing the "Next Col" button clears text boxes, resets the focus, and increments the block number so that the grid spacing for another column block can be entered. Type column block 2 first and last column entries in the appropriate text boxes. Choosing the "Enter Col" button stores column entries. Choosing the "Next Col" button clears text boxes, resets the focus, and increments the block number so that the grid spacing for another column block can be entered. Repeat this process until all column block grid spacings have been entered.

Choosing the "Row" button changes the block number to 1, indicating that data for row block 1 data can be entered, and resets the focus in the Block Row Data text box. Type row block number 1 first and last row numbers and the row grid spacing in the appropriate Block Row Data text boxes. Choosing the "Enter Row" button stores row numbers and the grid spacing. Choosing the "Next Row" button clears text boxes, resets the focus, and increments the block number so that the grid spacing for another row block can be entered. Type row block 2 first and last row numbers and the row grid spacing in the appropriate text boxes. Choosing the "Enter Row" button stores row entries. Choosing the "Next Row" button clears text boxes, resets the focus, and increments the block number so that the grid spacing for another row block can be entered. Repeat this process until all row block grid spacings have been entered. Type the interpolation grid (model cell nodes) upper left node X-coordinates, upper left node Y-coordinate, left column, right column, upper row, and lower row numbers in the appropriate text boxes at the bottom of the dialog box. Choosing the "Enter" button stores the interpolation grid data.

If "Surface" and "Triangulation" were selected, choosing the "OK" button in the Specify Grid Spacing Dimensions dialog box displays a Specify Point Dimensions data entry dialog box prompting for information about the scattered control points. Type the first and last point numbers in the appropriate text boxes. Point numbers 4–500 are supported. Choosing the "Enter" button stores these entries. Choosing the "OK" button stores entries, calculates interpolation grid values using the triangulation technique, closes the dialog box, and returns to the Manager.

If "Surface" and "Kriging" were selected, choosing the "OK" button in the Specify Grid Spacing Dimensions dialog box displays a Specify Point Dimensions data entry dialog box prompting for information about the scattered control points. Type the first and last point numbers in the appropriate text boxes. Point numbers 4–500 are supported. Choose one of two Kriging Output options: "Z Value" or "Error". Z Value refers to interpolation grid Z value estimates, and Error refers to interpolation grid standard errors of estimate. Choosing the "Enter" button stores these entries. Choosing the "OK" button stores entries, calculates interpolation grid Z values or standard errors of estimate using the kriging technique, closes the dialog box, and returns to the Manager.

Time Steps

Selecting "Time Step" displays a Specify Time Increment Dimensions data entry dialog box prompting for information about time discretization. Type the stress period number, number of time steps, time at end of stress period, time step multiplier, and representative cell grid spacing, transmissivity, storativity, and flow velocity in the appropriate text boxes. Stress period numbers 1–100 are supported. Choosing the "Enter" button stores these entries and calculates and displays the initial time increment, critical flow time increment, and critical transport time increment. Choosing the "Next" button clears text boxes and resets the focus so that data for additional stress periods can be entered. Choosing the "OK" button closes the dialog box and returns to the Manager.

VIEW MENU

Prior to selecting the View menu, an appropriate array must be preloaded or pre-edited. Selecting the "View" menu displays a drop-down menu with four options: Tabular, Map, Graph, and Help. Tabular refers to an array tabular (spreadsheet) display, Map refers to a Z value array zonal diagram for checking the logical consistency of the array or a Z value contour display, and Graph refers to an XY array chart display. Finite devices like screens and printers impose limits on the amount of data that can be meaningfully viewed at any one time. Multiple-block viewports are supported so that arrays can be viewed in their entirety or piecemeal by choosing proper block dimensions. Complicated maps can be divided into submaps so that realistic images can be produced on the screen or on paper. Positive site base map coordinates are supported. Choosing the "Cancel" button in any View dialog box ends the dialog before completion and returns to the Manager.

The graphics displayed in *DesignMod*'s Map and Graph options were designed to assist in the analysis of data and are draft quality. You should use commercial graphics software such as *GRAPHER* and *SURFER* (available from Golden Software, Inc., P.O. Box 281, Golden, CO 80402) for publication-quality graphics.

Tabular

Selecting "Tabular" displays a cascading menu with five options: Log, Spacing, Grid, Doublet, and Triplet. Log refers to a well log or model data array, Spacing refers to a column or row grid spacing array, Grid refers to a Z value array, Doublet refers to an XY array, and Triplet refers to a gridded or scattered XYZ array. Selecting "Log" displays a cascading menu with four options: Geology, Hydrogeology, Geochemistry, and Model. Geology refers to a well log array based on geologic units, Hydrogeology refers to a well log array based on hydrogeologic units, Geochemistry refers to a well log array based on hydrogeologic units and selected constituents, and Model refers to a model data array. Selecting "Triplet" displays a cascading menu with two options: Scattered and Gridded. If "Gridded" is selected, the number of columns multiplied by the number of rows must not exceed 1600. Selecting any one of the "Geology", "Hydrogeology", "Geochemistry", "Model", "Spacing", "Grid", "Doublet", "Scattered", or "Gridded" options results in the appearance of a cascading menu with two options: Screen and Printer. Screen refers to a screen display of the array and Printer refers to a hard copy of the array.

Selecting "Tabular" — "Log" — Geology", "Hydrogeology", "Geochemistry", or "Model" — "Screen" displays a Specify Tabular Display Dimensions data entry dialog box prompting for information about the viewport block number and the well or model whose data are to be viewed. Type the block number, well or model number, first unit or layer number, and last unit or layer number in appropriate text boxes. In addition, when "Geochemical" is selected, type the first constituent number and the last constituent number in the appropriate text boxes. Block and well or model numbers 1–100 are supported. If "Geology", "Hydrogeology", or "Model" was selected, unit or layer numbers 1–25 are supported. If "Geochemistry" was selected, unit or layer numbers 1–10 and constituent numbers 1–25 are supported. Choosing the "Enter" button stores these entries. Choosing the "Next" button clears text boxes and resets the focus so that data for additional blocks can be entered.

Choosing the "OK" button displays a Tabular Display spreadsheet and dialog box prompting for information about the viewport block and the display format. Type the block number associated with the well or model for which data are to be displayed in the appropriate text box. Specify the format to be used in displaying numerical data by selecting one of the two Format options: "Decimal" or "Exponential". The Decimal option should be selected when the number of value digits to the left of the decimal point exceeds five or the number of value digits to the right of the decimal point exceeds two. Choosing the "View" button displays array values in the spreadsheet box. Use the spreadsheet box scroll bars or arrow keys to view various portions of the array. Choosing the "Clip" button after highlighting appropriate spreadsheet cells places the contents of these cells onto the *Windows* Clipboard so they can be pasted into other applications. Choosing the "Next" button clears text boxes and resets the focus so that data for other wells or models can be viewed. Choosing the "OK" button closes the dialog box and returns to the Manager.

Selecting "Tabular" — "Log" — "Geology", "Hydrogeology", "Geochemistry", or "Model" — "Printer" displays a Specify Log Printer Dimensions data entry dialog box prompting for information about the well or model whose data are to be printed and the display format. Type the well or model number, first unit or layer number, last unit or layer number, X- and Y-coordinate unit, and elevation unit in appropriate text boxes. Specify the format to be used in displaying numeric data by selecting one of the two Format options: "Decimal" or "Exponential". The Decimal option should be selected when the number of value digits to the left of the decimal point exceeds five or the number of value digits to the right of the decimal point exceeds two. In addition, when "Geochemistry" is selected, type the first constituent number and the last constituent number in the appropriate text boxes. Well or model numbers 1–100 and 12 unit characters are supported. If "Geologic", "Hydrogeologic", or "Model" was selected, unit or layer numbers 1–25 are supported. If "Geochemical" was selected, unit or layer numbers 1–10 and constituent numbers 1–25 are supported. Choosing the "Print" button stores these entries, creates a print file, and sends the print file to the *Windows* Print Manager to obtain a hard copy of the data for the well or model. Choosing the "Next" button clears text boxes and resets the focus so that data for additional wells or models can be printed. Choosing the "OK" button closes the dialog box and returns to the Manager.

Selecting "Tabular", "Spacing", and "Screen" displays a Specify Grid Spacing Screen Dimensions data entry dialog box prompting for information about the column or row grid spacings to be viewed on the screen. Type the grid spacing block number and unit in the appropriate text boxes. Block numbers 1–100 are supported. Specify the data type by selecting one of two Data Type option buttons: "Column" or "Row". Type the first and last column or row numbers within the grid spacing block number in appropriate text boxes. Column and row numbers 1–200 are supported. Choosing the "Enter" button stores these entries. Choosing the "Next" button clears text boxes and resets the focus so that data for another grid spacing block can be entered. Choosing the "OK" button displays a Tabular Display grid and dialog box prompting for information about the grid spacing block to be viewed.

Type the block number associated with the grid spacing data to be displayed in the appropriate text box. Specify the format that will be used in displaying numeric data by selecting one of the two Format options: "Decimal" or "Exponential". The Decimal option should be selected when the number of value digits to the left of the decimal point exceeds five or the number of value digits to the right of the decimal point exceeds two. Choosing the "Facts" button displays a Fact Message box containing information about the grid spacing array. The array first and last column or row numbers and maximum and minimum column or row grid spacings are displayed in the message box. Choosing the Fact Message box "OK" button returns to the Specify Grid Spacing Screen Dimensions dialog box. Choosing the "View" button displays grid spacing values in the spreadsheet box. Use the spreadsheet box scroll bars or the arrow keys to view various portions of the array. Choosing the "Clip" button after highlighting appropriate spreadsheet box cells places the contents of these cells onto the *Windows* Clipboard so they can be pasted into other applications. Choosing the "Next" button clears text boxes and resets the focus so that additional blocks of the grid spacing array can be viewed. Choosing the "OK" button closes the dialog box and returns to the Manager.

Selecting "Tabular", "Spacing", and "Printer" displays a Specify Grid Spacing Printer Dimensions data entry dialog box prompting for information about the column or row grid spacings to be printed. Specify the data type by selecting one of the two Data Type option buttons: "Column Grid Spacing" or "Row Grid Spacing". Specify the format to be used in displaying numeric data by selecting one of the two Format options: "Decimal" or "Exponential". The Decimal option should be selected when the number of value digits to the left of the decimal point exceeds five or the number of value digits to the right of the decimal point exceeds two. Type the first and last column or row numbers within the specified block and the grid spacing unit in the appropriate text boxes. Column and row numbers 1–200 and 12 unit characters are supported. Choosing the "Print" button stores these entries, creates a print file, and sends the print file to the *Windows* Print Manager to obtain a hard copy of the block of grid spacing data. Choosing the "Next" button clears text boxes and resets the focus so that data for additional blocks can be printed. Choosing the "OK" button closes the dialog box and returns to the Manager.

Selecting "Tabular", "Grid", and "Screen" displays a Specify Tabular Screen Display Dimensions data entry dialog box prompting for information about the size of the grid block containing the Z values to be displayed on the screen. Type the block number in the appropriate text box. Block numbers 1–100 are supported. Select the "Z Value Display Type" option button. Type the upper left and lower right corner columns and rows for the specified block number and the array Z value unit in the appropriate text boxes. Column and row numbers 1–200 and 12 unit characters are supported. Entries must be within the dimensions of any preloaded or pre-edited grid or gridded triplet array or unexpected results can occur. Choosing the "Enter" button stores these entries. Choosing the "Next" button clears the text boxes and resets the focus so that data for additional blocks can be entered.

Choosing the "OK" button displays a Tabular Display spreadsheet and dialog box prompting for information about the block number and the format to be used in displaying Z values and displaying the Z value unit. Type the number of the block containing the Z values to be displayed in the appropriate text box. Specify the format to be used in displaying numeric data by selecting one of the two Format options: "Decimal" or "Exponential". The Decimal option should be selected when the number of value digits to the left of the decimal point exceeds five or the number of value digits to the right of the decimal point exceeds two. Choosing the "Facts" button displays a Fact Message box containing information about the grid array. The array upper left and lower right corner column or row numbers and maximum and minimum Z values and associated columns or rows are displayed in the message box. Choosing the Fact Message box "OK" button returns to the Tabular Display dialog box. Choosing the Tabular Display dialog box "View" button displays array Z values in the spreadsheet box. Use the spreadsheet box scroll bars or arrow keys to view various portions of the array. Choosing the "Clip" button after highlighting appropriate spreadsheet cells places the contents of these cells onto the *Windows* Clipboard so they can be pasted into other applications. Choosing the "Next" button clears text boxes and resets the focus so that additional blocks of the grid array can be viewed. Choosing the "OK" button returns to the Manager.

Selecting "Tabular", Grid", and "Printer" displays a Specify Printer Block Dimensions data entry dialog box prompting for information about the size of the grid block containing the Z values to be printed and the format to be used in displaying the Z values. Data entries must be within the dimensions of the preloaded or pre-edited grid or gridded triplet array or unexpected results can occur. Type the grid block upper left and lower right corner columns and rows and the Z value unit in the appropriate text boxes. Column and row numbers 1–200 and 12 unit characters are supported. Specify the format to be used in displaying numeric data by selecting one of the two Format options: "Decimal" or "Exponential". The Decimal option should be selected when the number of value digits to the left of the decimal point exceeds five or the number of value digits to the right of the decimal point exceeds two. Choosing the "Print" button stores these entries, creates a print file, and sends the print file to the *Windows* Print Manager to produce a hard copy of the block Z values. Choosing the "Next" button clears text boxes and resets the focus so that Z values for additional grid array blocks can be printed. Choosing the "OK" button returns to the Manager.

Choosing "Tabular", "Doublet", and "Screen" displays a Specify Tabular Display Dimensions data entry dialog box prompting for information about the size of XY array block to be viewed. Type the block number, the first and last point numbers for the block number, and the X- and Y-coordinate unit in the appropriate text boxes. Block numbers 1–100, first point and last point numbers 1–500, and 12 X- and Y-coordinate unit characters are supported. Entries must be within the dimensions of the preloaded or pre-edited doublet array or unexpected results can occur. Choosing the "Enter" button stores these entries. Choosing the "Next" button clears text boxes and resets the focus so that values for additional point blocks can be entered.

Choosing the "OK" button displays a Tabular Display spreadsheet and dialog box prompting for information about the block to be viewed and the format to be used in displaying XY values. Type the number of the block containing the data to be viewed in the appropriate text box. Specify the format to be used to display numeric data by selecting one of the two Format options: "Decimal" or "Exponential". The Decimal option should be selected when the number of value digits to the left of the decimal point exceeds five or the number of value digits to the right of the decimal point exceeds two. Choosing the "Facts" button displays a Fact Message box containing information about the doublet array. The array first and last point numbers and maximum and minimum X and Y values and associated points are displayed in the message box. Choosing the Fact Message box "OK" button returns to the Tabular Display dialog box. Choosing the "View" button displays array XY values in the spreadsheet box. Use the spreadsheet box scroll bars or arrow keys to view various portions of the array. Choosing the "Clip" button after highlighting appropriate spreadsheet cells places the contents of these cells onto the *Windows* Clipboard so they can be pasted into other applications. Choosing the "Next" button clears the Block Number text box and resets the focus so that additional blocks of XY values can be viewed. Choosing the "OK" button returns to the Manager.

Selecting "Tabular", "Doublet", and "Printer" displays a Specify Tabular Display Dimensions data entry dialog box prompting for information about the size of the XY array block to be printed. Entries must be within the dimensions of the preloaded or pre-edited doublet array or unexpected results can occur. Type the first and last point block numbers and X- and Y-coordinate units in the appropriate text boxes. First point and last point numbers 1–500 and 12 X- and Y-coordinate unit characters are supported. Specify the format to be used to display numeric data by selecting one of the two Format options: "Decimal" or "Exponential". The Decimal option should be selected when the number of value digits to the left of the decimal point exceeds five or the number of value digits to the right of the decimal point exceeds two. Choosing the "Print" button stores these entries, creates a print file, and sends the print file to the *Windows* Print Manager to produce a hard copy of the block XY values. Choosing the "Next" button clears text boxes and resets the focus so that values for additional XY array blocks can be printed. Choosing the "OK" button returns to the Manager.

Selecting "Tabular", "Triplet", "Scattered", and "Screen" displays a Specify Tabular Display Dimensions data entry dialog box prompting for information about the size of the scattered triplet array whose XYZ values are to be viewed. Type the number of the block containing the data to be viewed, the first and last point numbers for the block number, and the X- and Y-coordinate and Z value units in the appropriate text boxes. Block numbers 1–100, first point and last point numbers 1–500, and 12 X- and Y-coordinate and Z value unit characters are supported. Entries must be within the dimensions of the preloaded or pre-edited gridded triplet array or unexpected results can occur. Choosing the "Enter" button stores these entries. Choosing the "Next" button clears text boxes and resets the focus so that data for additional blocks can be entered.

Choosing the "OK" button displays a Tabular Display spreadsheet and dialog box prompting for information about the block to be viewed and the format to be used in displaying XYZ values. Type the number of the block containing the data to be viewed in the appropriate text box. Specify the format to be used to display numeric data by selecting one of the two Format options: "Decimal" or "Exponential". The Decimal option should be selected when the number of value digits to the left of the decimal point exceeds five or the number of value digits to the right of the decimal point exceeds two. Choosing the "Facts" button displays a Fact Message box containing information about the scattered triplet array. The array first and last point numbers and maximum and minimum X, Y, and Z values and associated points are displayed in the message box. Choosing the Fact Message Box "OK" button returns to the Tabular Display Dimensions dialog box. Choosing the Tabular Display Dimensions dialog box "View" button displays array XYZ values in the spreadsheet box. Use the spreadsheet box scroll bars or the arrow keys to view various portions of the array. Choosing the "Clip" button after highlighting appropriate spreadsheet cells places the contents of these cells onto the *Windows*

Clipboard so they can be pasted into other applications. Choosing the "Next" button clears the Block Number text box and resets the focus so that additional blocks of XYZ values can be viewed. Choosing the "OK" button returns to the Manager.

Selecting "Tabular", "Triplet", "Scattered", and "Printer" displays a Specify Tabular Display Dimensions data entry dialog box prompting for information about the size of the XYZ array block to be printed. Entries must be within the dimensions of the preloaded or pre-edited scattered triplet array or unexpected results can occur. Type the first and last point block numbers, X- and Y-coordinate unit, and Z value unit in the appropriate text boxes. First point and last point numbers 1–500 and 12 X- and Y-coordinate and Z value unit characters are supported. Specify the format to be used to display numeric data by selecting one of the two Format options: "Decimal" or "Exponential". The Decimal option should be selected when the number of value digits to the left of the decimal point exceeds five or the number of value digits to the right of the decimal point exceeds two. Choosing the "Print" button stores these entries, creates a print file, and sends the print file to the *Windows* Print Manager to produce a hard copy of the block XYZ values. Choosing the "Next" button clears text boxes and resets the focus so that values for additional XYZ array blocks can be printed. Choosing the "OK" button returns to the Manager.

Selecting "Tabular", "Triplet", "Gridded", and "Screen" displays a Specify Tabular Display Dimensions data entry dialog box prompting for information about the size of the grid block containing the XYZ values to be displayed on the screen. Type the number of the block containing the data to be viewed in the appropriate text box. Block numbers 1–100 are supported. Select the XYZ value "Display Type" option button. Type the upper left and lower right corner columns and rows for the block number, X- and Y-coordinate unit, and the array Z value unit in the appropriate text boxes. Column and row numbers 1–200 are supported. The number of columns multiplied by the number of rows must not exceed 1600. A total of 12 X- and Y-coordinate and Z value unit characters are supported. Type the upper left node X- and Y-coordinates in the appropriate text boxes. Entries must be within the dimensions of the preloaded or pre-edited gridded triplet array or unexpected results can occur. Choosing the "Enter" button stores these entries. Choosing the "Next" button clears the text boxes and resets the focus so that data for additional blocks can be entered.

Choosing the "OK" button displays the Tabular Display spreadsheet and dialog box prompting for information about the block number and the format to be used in displaying XYZ values. Type the number of the block containing the data to be viewed in the appropriate text box. Specify the format to be used to display numeric data by selecting one of the two Format options: "Decimal" or "Exponential". The Decimal option should be selected when the number of value digits to the left of the decimal point exceeds five or the number of value digits to the right of the decimal point exceeds two. Choosing the "Facts" button displays a Fact Message box containing information about the array.

The array upper left and lower right corner column or row numbers and maximum and minimum XYZ values and associated columns or rows are displayed in the message box. Choosing the Fact Message box "OK" button returns to the Tabular Display dialog box. Choosing the Tabular Display dialog box "View" button displays array XYZ values in the spreadsheet box. Use the spreadsheet box scroll bars or the arrow keys to view various portions of the array. Choosing the "Clip" button after highlighting appropriate spreadsheet cells places the contents of these cells onto the *Windows* Clipboard so they can be pasted into other applications. Choosing the "Next" button clears text boxes and resets the focus so that additional blocks of the gridded triplet array can be viewed. Choosing the "OK" button returns to the Manager.

Selecting "Tabular", "Gridded", and "Printer" displays a Specify Array Type dialog box. Specify the array type to be printed by selecting one of the two options: "Z Values" or "XYZ Values". Choosing the "OK" button stores the array type flag, closes the dialog box, and displays a Specify Printer Block Dimensions data entry dialog box prompting for information about the size of the grid block containing the XYZ values to be printed and the format to be used in displaying the Z values. Data entries must be within the dimensions of the preloaded or pre-edited gridded triplet array or unexpected results can occur. Type the number of the block containing the data to be printed, block upper left and lower right corner columns and rows, X- and Y-coordinate unit, and the Z value unit in appropriate text boxes. Column and row numbers 1–200 are supported. Be aware that large numbers of paper sheets will be used if you enter large numbers of columns and rows. Also, long periods of time will be required to print data for large numbers of columns and rows. A total of 12 X- and Y-coordinate unit and Z value unit characters are supported. Specify the format to be used to display numeric data by selecting one of the two Format options: "Decimal" or "Exponential". The Decimal option should be selected when the number of value digits to the left of the decimal point exceeds five or the number of value digits to the right of the decimal point exceeds two. Choosing the "Print" button stores these entries, creates a print file, and sends the print file to the *Windows* Print Manager to produce a hard copy of the block Z or XYZ values. Choosing the "Next" button clears text boxes and resets the focus so that values for additional gridded triplet array blocks can be printed. Choosing the "OK" button returns to the Manager.

Map

Contouring algorithms used in *DesignMod* consists of the detection of contour lines intersecting the four sides of screen pixel size grids (see Walton, 1991, pp. 287–288). Two grid-node Z values along a grid side are compared with the value of a contour. If the contour value lies between the grid-node values, the point location of the crossing is calculated with linear interpolation. A screen pixel is turned on at the point location of the contour. With sufficiently fine small grid sizes, turned-on pixels representing point contour locations merge together to look like a continuous curve. Other contouring algorithms are presented by Anand (1993).

Selecting "Map" displays a cascading menu with two options: Zonal Diagram and Contour. Selecting either of these options displays a cascading menu with two options: Screen and Printer. Selecting "Map", "Zonal Diagram", and "Screen" displays a Specify View Block Dimensions data entry dialog box prompting for information about the size of the block to be viewed and displaying the block number. Block numbers 1–100 are supported. Type the displayed block number upper left and lower right grid corner columns and rows, upper left column and row grid spacings, X- and Y-coordinate unit, and Z value unit for the displayed block number in appropriate text boxes. Column and row numbers 1–200 and 12 unit characters are supported. Entries must be within the dimensions of the preloaded or pre-edited gridded triplet array or unexpected results can occur. The upper left column and row grid spacing dimensions are entered so that grid spacings throughout the grid can be calculated with the gridded triplet array's X- and Y-coordinates.

Choosing the "Enter" button stores entries. Choosing the "Next" button increments the block number, clears text boxes, and resets the focus so that data for additional blocks can be entered. Choosing the "OK" button displays a Zonal Diagram Display picture and dialog box prompting for the block number and displaying information about the X- and Y-coordinate unit, the Z value unit, and the range number. Range numbers 1–12 are supported. Type the number of the block containing the values to be viewed in the appropriate text box. Block numbers 1–100 are supported. Type the lower ($\geq$) range value and upper ($<$) range value corresponding to the displayed range number in the appropriate text boxes. Choosing the "Enter" button stores these entries and displays the block upper left column and row, coordinates lower right column and row coordinates, and maximum and minimum Z values and corresponding column and row numbers in the upper left corner of the picture box. These coordinates disappear when the color zonal diagram is displayed. Choosing the "Next" button increments the range number, clears the range value text boxes, and resets the focus so that additional range values can be entered.

Choosing the "Plot" button displays grid blocks with colors in the picture box containing cell nodes whose Z values are within the range specified for colors. Unexpected results will occur if a gridded triplet array file with an upper left origin is preloaded using the lower left origin option. Choosing "Grid" displays a cell border line grid overlay. Selecting the "CR Or XY Axes" option displays either the column and row or XY coordinates of the mouse cross-hair to the right of the picture box when the left button of the mouse is depressed. The mouse cross-hair pointer senses its position relative to the upper left origin (0,0) of the picture box with a medium resolution, depending upon the picture box scale and your ability to center the cross-hair on an object. Positioning the cross-hair of the mouse on any block color after choosing the "Legend" button displays the range values corresponding to the color at the top of the screen. Choosing the "OK" button returns to the Manager.

Selecting "Zonal Diagram" and "Printer" displays a Specify View Block Diagram Dimensions data entry dialog box prompting for information about the size of the zonal diagram to be printed and displaying the block number. Block numbers 1–100 are supported. Type the printer block upper left and lower right grid corner column and row numbers and the upper left column and row grid spacings in the appropriate text boxes. Column and row numbers 1–200 are supported. Entries must be within the dimensions of the preloaded or pre-edited gridded triplet array or unexpected results can occur. Choosing the "Enter" button stores these entries. Choosing the "Next" button increments the block number, clears text boxes, and resets the focus so that data for additional blocks can be entered.

Choosing the "OK" button closes the dialog box and displays a Specify Printer Zonal Diagram Dimensions data entry dialog box prompting for information about the Z value ranges and displaying the range number. Type the number of the block containing the data to be printed in the appropriate text box. Block numbers 1–100 are supported. Type the lower and upper range values corresponding to the displayed range number and the Z value unit in the appropriate text boxes. A total of 12 Z value unit characters are supported. Choosing the "Unit" button stores the block number and the Z value unit entry. Choosing the "Range" button stores range values. Choosing the "Next Range" button increments the range number, clears the range lower and upper value text boxes, and resets the focus so that additional range data can be entered.

Choosing the "Print" button identifies cells whose Z values are within specified range values, creates a print file, and sends the print file to the *Windows* Print Manager so that a hard copy of the zonal diagram can be produced. Printed upper case letters represent cell nodes whose Z values are within the range values specified for the letter. The block upper left and lower right grid corner column and row numbers, upper left and lower right column and row coordinates, Z value unit, and the range value symbol legend also are printed. Choosing the "Next Block" button changes the range number, clears the range lower and upper value text boxes, and resets the focus so that range data for additional blocks can be entered. Choosing the "OK" button returns to the Manager.

Selecting "Map", "Contour", and "Screen" displays a Specify View Block Dimensions data entry dialog box prompting for information about the size of the contour map block and displaying the block number. Block numbers 1–100 are supported. Type the block upper left and lower right grid corner column and row numbers, upper left node column grid spacings, upper left node row grid spacing, X- and Y-coordinate unit, and Z value unit in the appropriate text boxes. Column and row numbers 1–200 are supported. A total of 12 X- and Y-coordinate unit and Z value unit characters are supported. Entries must be within the dimensions of the preloaded or pre-edited gridded triplet array or unexpected results can occur. The upper left node column grid spacing and the upper left node row grid spacing dimensions are entered so that grid spacings throughout the grid can be calculated with the gridded triplet array X- and Y-coordinates. Choosing the "Enter" button stores entries. Choosing the "Next" button increments the block number, clears text boxes, and resets the focus so that data for additional blocks can be entered.

Choosing the "OK" button calculates grid spacings and displays a Contour Map Display picture and dialog box. The dialog box prompts for information about the block number, displays the contour number, and displays the X- and Y-coordinate unit and Z value unit. Type the number of the block containing the data to be viewed and the contour value corresponding to the displayed contour number in the appropriate text boxes. Block numbers 1–100 and contour numbers 1–12 are supported. Choosing the "Enter" button stores these entries and displays the block upper left column and row numbers, lower right column and row numbers, and maximum and minimum Z values and corresponding column and row numbers in the picture box. This information disappears when contours are displayed. Choosing the "Next" button increments the contour number, clears the contour value text box, and resets the focus so that additional contour values can be entered.

Choosing the "Plot" button displays contours (curves connecting points of equal Z value in an XY plane) with colors and cell nodes in the picture box. Unexpected results will occur if a gridded triplet array file with an upper left origin is preloaded using the lower left origin option. Choosing "Grid" displays a cell border line grid overlay. Be aware that the calculation time will be long if the grid contains a large number of columns and rows. Selecting the "CR Or XY Axes" option displays either the column and row or XY coordinates of the mouse cross-hair to the right of the picture box when the left button of the mouse is depressed. The mouse cross-hair pointer senses its positive relative to the upper left origin (0,0) of the picture box with a medium resolution, depending upon the picture box scale and your ability to center the cross-hair on an object. Positioning the mouse cross-hair on a color after choosing the "Legend" button displays the contour value corresponding to the color at the top of the screen. Choosing the "Block" button clears text boxes and resets the focus so that contours within additional blocks can be viewed. Choosing the "OK" button returns to the Manager.

Selecting "Map", "Contour", and "Printer" displays a Specify View Block Dimensions data entry dialog box prompting for information about the size of the block to be printed and displaying the block number. Block numbers 1–100 are supported. Type the block upper left and lower right grid column and row numbers and upper left column and row grid spacings in the appropriate text boxes. Column and row numbers 1–200 are supported. Entries must be within the dimensions of the preloaded or pre-edited gridded triplet array or unexpected results can occur. The block upper left column grid spacing and the block upper left row grid spacing dimensions are specified so that grid spacings throughout the grid can be calculated with the gridded triplet array X- and Y-coordinates. Choosing the "Enter" button stores entries. Choosing the "Next" button increments the block number, clears text boxes, and resets the focus so that data for another block can be entered.

Choosing the "OK" button displays a Specify Printer Contour Map Dimensions data entry dialog box prompting for information about contour values and displaying the contour number. Type the number of the block containing the data to be printed in the appropriate text box. Type the X- and Y-coordinate unit and Z value unit in the appropriate text boxes. Block numbers 1–100 and 12 unit characters are supported. Choosing the "Unit" button stores these entries. Type the contour value corresponding to the displayed contour number in the text box. Choosing the "Contour" button stores this entry. Choosing the "Next Contour" button increments the contour number, clears the contour value text boxes, and resets the focus so that additional contour values can be entered. Contour numbers 1–12 are supported. Choosing the "Print" button calculates contour X- and Y-coordinates, creates a print file, and sends the print file to the *Windows* Print Manager so that a hard copy of the contour map can be produced. Be aware that the calculation time will be long if the grid contains a large number of columns and rows. Z value surface contours are printed within a box at the top of the page with the block upper left and lower right grid column and row numbers and row X- and Y-coordinates. The X- and Y-coordinate unit, Z value unit, and the contour value label legend are printed below the box. Choosing the "Next Block" button changes the block number, clears text boxes, and resets the focus so that data for another block can be printed. Choosing the "OK" button returns to the Manager.

Graph

Selecting "Graph" displays a cascading menu with three options: Arithmetic, Semilog, and Logarithmic. Arithmetic refers to an XY graph with an arithmetic X-axis and an arithmetic Y-axis; Semilog refers to an XY graph with a logarithmic X-axis and an arithmetic Y-axis; and Logarithmic refers to an XY graph with a logarithmic X-axis and a logarithmic Y-axis. Selecting any one of these three options results in the appearance of a cascading menu with two options: Screen and Printer. Screen refers to a screen view of the graph, and Printer refers to a hard copy of the graph.

Selecting "Graph", "Arithmetic", and "Screen"; "Graph", "Semilog", and "Screen"; or "Graph", "Logarithmic", and "Screen" results in the appearance of a Specify Screen Graph Dimensions data entry dialog box prompting for information about the graph features. Type the X-axis start value; X-axis end value; Y-axis start value; Y-axis end value; X-axis grid spacing and Y-axis grid spacing (if Arithmetic was selected) or X-axis log cycles and Y-axis grid spacing (if Semilog was selected) or X-axis log cycles and Y-axis log cycles (if Logarithmic was selected); X-axis label and unit; and Y-axis label and unit in the appropriate text boxes. Axis start and end values define the axis grid frame. Axis grid spacings define spaces between axis grid lines. Log cycles refers to the number of log cycles on an axis; 10 log cycles and 20 label and unit characters are supported. If "Semilog" or "Logarithmic" was selected, X-axis start and end values must be 1 raised to some power of 10. If "Logarithmic" was selected, Y-axis start and end values must be 1 raised to some power of 10. Choosing the "Enter" button stores entries. Choosing the "OK" button displays a Graph Display picture and data entry dialog box.

The Graph Display picture and data entry dialog box prompts for information about the graph origin and the type of display. An upper left or lower left graph origin is specified by selecting one of the two Origin option buttons. Whether point symbols or lines connecting points are to be displayed is specified by selecting one of the two Display option buttons. The X-axis and Y-axis labels and units are displayed below the option buttons. The X- and Y-coordinates of the mouse pointer are displayed below the labels and units when the left button of the mouse is depressed. Choosing the "Plot" button stores entries and displays the graph in the picture box with point symbols or lines between points and axis grid line intersections. Choosing the "Grid" button displays axis grid frame overlays. Choosing the "OK" button closes the dialog box and returns to the Manager.

Selecting "Graph", "Arithmetic", and "Printer"; "Graph", "Semilog", and "Printer"; or "Graph", "Logarithmic", and "Printer" displays a Specify Printer Graph Dimensions data entry dialog box prompting for information about graph features. Type the X-axis start value; X-axis end value; Y-axis start value; Y-axis end value; X-axis grid spacing and Y-axis grid spacing (if Arithmetic was selected) or X-axis log cycles and Y-axis grid spacing (if Semilog was selected) or X-axis log cycles and Y-axis log cycles (if Logarithmic was selected); X-axis label and unit; and Y-axis label and unit in the appropriate text boxes.

An upper left or lower left graph origin is specified by selecting one of two Origin option buttons. Point or line display is specified by selecting one of the two Display option buttons. Axis line intersections or grid frames are specified by selecting one of the two Axis option buttons. Axis start and end values define the axis grid frame. Axis grid spacings define spaces between axis grid lines. Log cycles refers to the number of log cycles on an axis; 10 log cycles and 20 label and unit characters are supported. Choosing the "Enter" button stores entries and option choices. Choosing the "Print" button calculates graph X- and Y-coordinates, creates a print file, and sends the print file to the *Windows* Print Manager so that a hard copy of the graph can be produced. The graph is printed at the top of the page with the axis start and end values and X- and Y-axis labels and units. Choosing the "OK" button closes the dialog box and returns to the Manager.

DRAW MENU

Draw provides an easy-to-use onscreen cross section or plan view drawing pad (Design Window). It enables you to draw to scale and edit site base map objects (shapes) such as rectangles, circles, ellipses, and arcs and their component entities such as points and lines and to label the drawing with text. Definition points of objects and entities, such as the opposite corners of a rectangle, the center of a circle and a point on its circumference, and the beginning and ending points of a line, are defined with the keyboard or mouse. Positive site base map coordinates are supported.

To avoid clutter and overlapping, a complex, complicated, or large modeling area of interest can be divided into sections (viewport blocks) so that, when organized, each one shows a different portion of the modeling area of interest without clutter. You can easily return to any particular viewport block. Block dimensions set the size of the on-screen viewport to suit your needs. The upper left and lower right corner dimensions of the viewport block are used to identify the portion of the modeling area of interest that will be displayed.

Appropriate geologic, hydrogeologic, geochemical, and/or model log arrays must be imported, created, or edited before attempting to draw cross sections or plan views. Selecting the Draw menu displays a drop-down menu with two options: Cross Section and Plan View. Choosing "Draw" and "Cross Section" displays a Specify Cross Section Dimensions data entry dialog box prompting for information about the vertical and horizontal dimensions of the cross section and displaying the viewport block number. Block numbers 1–100 are supported. Specify the type of cross section by selecting one of the two Cross Section Type options: Well or Model. If "Well" was selected, type the upper and lower cross section vertical elevations (leave space for well numbers to be printed on top of the log), number of wells on the cross section block, user's number for each well on the cross section block, and cross section vertical and horizontal units in appropriate text boxes. Vertical and horizontal dimension units each can have no more than 12 characters. Choosing the "Enter Well" button stores entries. Choosing the "Next Well" button clears the Well Number text box and resets the focus so that data for additional wells can be entered. If the "Model" option was selected, type the upper and lower cross section elevations and the cross section vertical unit in appropriate text boxes. Choosing the "Enter Block" button stores entries for the block. Choosing the "Next Block" button increments the cross section viewport block number, clears text boxes, and resets the focus so that data for additional cross section viewport blocks can be entered. Choosing the "OK" button closes the dialog box and displays the Cross Section Design Window. The following custom modeling tools along the left side of the picture box are enabled: Model or Log (depending on the selected cross section type), Undo, Save, Load, Help, OK, and Cancel.

Choosing "Draw" and "Plan View" displays a Specify Plan View Dimensions data entry dialog box prompting for information about the plan view areal extent and displaying the viewport block number. Block numbers 1–100 are supported. Site base map Cartesian coordinates are supported, measured from a lower left origin. The first trial model grid axes are parallel to the site base map coordinate axes or are rotated counterclockwise or counterclockwise about the site map X-axis. Type the upper left and lower right first trial model grid corner site base map X- and Y-coordinates, X- and Y-coordinate units, model grid rotation angle, and X- and Y-axis offsets in appropriate text boxes. The upper left and lower right first trial model grid corner site base map X- and Y-coordinates are measured from the site base map origin. The upper left first trial model grid corner Y-coordinate and the lower right first trial model X-coordinate must exceed 0. Whole first trial model grid rotation angles of 0–90 degrees are supported. Offsets are measured between the lower left first trial model grid corner and the site base map lower left origin. Choosing the "Enter" button stores entries. Choosing the "Next" button increments the block number, clears text boxes, and resets the focus so that data for additional blocks can be entered. Choosing the "OK" button closes the dialog box and displays the Plan View Design Window. The following custom modeling tools are enabled: Grid, Undo, Save, Load, Help, OK, and Cancel.

The Cross Section and Plan View Design Windows contain a drawing area (picture box) in the center with a mouse cross-hair cursor surrounded by Toolboxes running down the right and left sides of the picture box, a viewport block number text box centered at top of the picture box, an X- and Y-coordinate display above and to the left of the picture box (i.e., displays how far the cursor is from the picture box upper left origin), a column and row display above and to the right of the picture box, a unit display below and to the left of the picture box, and a legend display below and to the right of the picture box. The Toolbox on the right offers an assortment of primitive objects and style, pattern, and color settings (Point, Line, Box, Poly, Circle, Ellipse, Arc, Curve, Text, Font, Style, Color, Width, Fill, Erase, and Key) to help you draw site base map features in the picture box. That Toolbox also enables you to type characters and numbers on your drawing; erase text, objects, and entities; and select the drawing mode (Mouse or Keyboard). The Toolbox on the left offers an assortment of custom modeling tools (Model, Log, WL (water level), Chem, Undo, Info, Cell, Zone, Map, Legend, Save, Load, Help, OK, and Cancel) to help you draw and erase model grids, well logs, data points, model grid cells, model grid cell zones, contour and zonation maps, and legends. That Toolbox also enables you to close or cancel the Design Window and to obtain help. The top center Next button can be used to clear the picture box and change the viewport block number.

Following instructions at the top of the picture box, type the desired viewport block number in the top center text box and choose the top center

"Enter" button to store that number and set the picture box scale. The vertical and horizontal scales of the picture box are set consistent with the previously specified vertical elevations and the X- and Y-coordinates of the selected wells, respectively, if a cross section is to be viewed. The Cross Section Design Window has a lower left origin (0,Z). X-distances from the beginning of the cross section increase from left to right, and Z-coordinates increase from bottom to top. The Plan View Design Window has an upper left origin (0,0). X-coordinates and model grid column numbers increase from left to right, and Y-coordinates and model grid row numbers increase from top to bottom. The plan view coordinate system coincides with the modeling finite-difference coordinate system but is opposite to the site base map Cartesian coordinate system, which has a lower left origin with X-coordinates increasing from left to right and Y-coordinates increasing from bottom to top. Cartesian site base map coordinates are transformed (offset and rotated) to the model finite-difference grid coordinate system before being used to draw objects or type text (see Rumbaugh, 1989, pp. 689–691).

The cross-hair pointer of the mouse senses its positive relative to the upper left origin (0,0) of the picture box with a medium resolution. The motion of the mouse sends a continuous stream of coordinates that indicate the position of the cross-hair in the X, Y or Z column and row formats. The mouse does not have the resolution (1000 parts per inch) of a digitizer tablet and stylus or puck used in CAD programs. Reported coordinates are slightly in error depending upon the picture box scale, line width, and your ability to center the cross-hair on an object.

Before drawing site base map objects and entities or typing text in the picture box, you may want to change line and text default settings, which are:Width (1 pixel), Style (solid), Interior Fill (transparent), Color (black), Font Size (8.25), Font Name (Helvetica), and Font Color (black). This can be accomplished by choosing the Width, Style, Fill, Color, and/or Font tools. Changes affect only the single object or text you draw immediately after changing default values, not the objects and text currently existing in your drawing. Default values are reset after each object or text is drawn. Choose "Color" before choosing Width, Style, Fill, or Font; color affects text, the outline or frame of a shape, and the fill you draw. To select the tool you want to use, point to the tool button and click the mouse button, or hold down Alt and press the underlined letter in the tool button caption, or use the Tab key.

Choosing "Color" displays a Select Color dialog box offering a palette of 16 colors: Black, Blue, Green, Cyan, Red, Magenta, Yellow, White, Gray, Light Blue, Light Green, Light Cyan, Light Red, Light Magenta, Light Yellow, and Bright White. Select a color from the Color List box using the scroll bars or arrow keys, thereby displaying a sample color block. Choosing the "OK" button stores the color selection flag and closes the dialog box.

Choosing "Width" displays a Select Line Width dialog box offering a selection of five line widths: 1 pixel, 2 pixels, 3 pixels, 4 pixels, and 5 pixels.

Select a line width from the Width list box using the scroll bars or arrow keys, thereby displaying a sample of the line width. If the selected width is greater than 1 pixel, you lose control of the line style and the line is solid regardless of the width. Choosing the "OK" button stores the line width selection flag and closes the dialog box.

Choosing "Style" displays a Select Line Style dialog box offering a selection of seven line styles: solid, dash, dot, dash-dot, dash-dot-dot, invisible, and inside solid. Select a line style from the Style List box using the scroll bars or arrow keys, thereby displaying a sample of the line style. Choosing the "OK" button stores the style selection flag and closes the dialog box.

Choosing "Fill" displays a Select Fill Style dialog box offering a selection of eight fill styles: transparent, solid, horizontal lines, vertical lines, up diagonal lines, down diagonal lines, cross lines, and diagonal cross lines. Select a fill style from the Fill list box using the scroll bars or arrow keys, thereby displaying a sample of the fill style. Choosing the "OK" button stores the fill selection flag and closes the dialog box.

Choosing "Font" displays a Specify Font Attributes dialog box offering a selecting of three font names, Helv (Helvetica), Roman, and Modern; six font sizes, 8.25, 9.75, 12, 13.5, 18, and 24; and a palette of 16 colors, Black, Blue, Green, Cyan, Red, Magenta, Yellow, White, Gray, Light Blue, Light Green, Light Cyan, Light Red, Light Magenta, Light Yellow, and Bright White. Select a font name, size, and color from the Name, Size, and Color List boxes using the scroll bars or arrow keys. Choosing "Enter" displays a sample text block. Choosing the "OK" button stores font selection flags and closes the dialog box.

Choosing "Key" displays a Select Drawing Mode dialog box offering two drawing mode options: Keyboard and Mouse. Keyboard refers to drawing objects with keyboard coordinate and other data entry. Mouse refers to drawing objects (except ellipse, arc, and curve) with the mouse and without keyboard coordinate and other data entry. Select the desired drawing mode and choose the "OK" button to close the dialog box.

In the Mouse drawing mode, to draw a point in the picture box choose the "Point" tool, position the cross-hair where you want a dot to appear, and click the mouse button. To draw a line from one point to another, choose the "Line" tool, position the cross-hair where you want the line to start, hold the mouse button down, and drag the mouse. When the line is the right length, release the mouse button.

In the Mouse drawing mode, objects are drawn from one corner to the opposite corner outward. If you position the cross-hair precisely at any picture box border and attempt to draw an object, unexpected results will occur. In drawing a rectangle, choose the "Box" tool, position the cross-hair at one of the object's corners, hold down the mouse button, and drag diagonally away from the starting point. When the rectangle is the right size, release the mouse button. To draw a circle from its center outward, choose the "Circle" tool, position the cross-hair at the point where you want the circle center, hold down the mouse button, and drag outward. To draw irregular shapes or closed objects, choose the "Poly" tool and draw the first line of the polygon. With the cross-hair at the spot where the first line ends and the second line begins, click the mouse button. Continue this procedure until the polygon is completed. To erase (remove) objects, entities, or text from a rectangular area, choose the "Erase" tool, position the cross-hair at one end of the rectangular area to be erased, hold down the mouse button, and drag diagonally away from the starting point. When the rectangle is the right size, release the mouse button. Ellipse, arc, and curved objects and entities are drawn with keyboard data entry as explained later. To add text to your drawing, position the cross-hair at the spot where the upper left corner of the text string is to start and click the mouse button. Choose the "Text" tool and type. In the Mouse mode, *DesignMod* supports only horizontal single-line text strings up to 255 characters and produces plain text without bold, italic, or underlined attributes using the Helvetica font, black color, and 8.25 size.

The keyboard mode assumes that you have obtained object and entity definition point coordinates by digitizing site map coordinates or reading model coordinates displayed above and to the left of the picture box. Choosing any object or entity tool displays a dialog box wherein you select a Coordinate Type option button (Screen or Site). Screen refers to model X- and Y- or X- and Z-coordinates with upper left origin as displayed above and to the left of the picture box. Site refers to X- and Y- or X- and Z-coordinates from a site base map with lower left origin.

In the Keyboard mode, choose the "Point" tool to draw a point in the picture box. A Specify Point Dimensions dialog box appears prompting for information about the point coordinates. Specify the type of line coordinates you are going to enter by selecting one of the two Coordinate Type options: "Screen" or "Site". Type the screen or site coordinates in the appropriate text boxes. The point is drawn in the picture box when you choose the "Draw" button. Choosing "Next" clears the text boxes and resets the focus so that additional points can be drawn. Choosing the "OK" button closes the dialog box.

In the Keyboard mode, choose the "Line" tool to draw a line in the picture box. The line can represent such features as the screened interval or water level in a well, part of a vertical and/or horizontal scale, or part of a north arrow. A Specify Line Dimensions dialog box appears prompting for information about the line definition points. Specify the type of line coordinates you are going to enter by selecting one of the two Coordinate Type options: "Screen" or "Site". Type the screen or site start and end of line coordinates in the appropriate text boxes. The line is drawn in the picture box when you choose the "Draw" button. Choosing "Next" clears the text boxes and resets the focus so that additional lines can be drawn. Choosing the "OK" button closes the dialog box.

In the Keyboard mode, choose the "Box" tool to draw a box in the picture box. A Specify Box Dimensions dialog box appears prompting for information about the box definition points. Specify the type of box coordinates you are going to enter by selecting one of the two Coordinate Type options: "Screen" or "Site". Type the screen or site upper left and lower right corner box coordinates in the appropriate text boxes. The box is drawn in the picture box when you choose the "Draw" button. Choosing "Next" clears the text boxes and resets the focus so that additional boxes can be drawn. Choosing the "OK" button closes the dialog box.

In the Keyboard mode, choose the "Poly" tool to draw a polygon in the picture box. A Specify Polygon Dimensions dialog box appears prompting for information about the polygon definition points. Specify the type of polygon coordinates you are going to enter by selecting one of the two Coordinate Type options: "Screen" or "Site". Type the number of vertices in the polygon and the screen or site vertice coordinates in the appropriate text boxes. The number of vertices must be at least 6 and cannot exceed 100. Choosing the "Enter" button stores coordinates for the vertice whose number is displayed. Vertices are numbered and vertice coordinates must be entered consecutively proceeding in a clockwise direction around the polygon. Choosing "Next" increments the vertice number, clears text boxes, and resets the focus so that you can enter data for additional vertices. After coordinates for all vertices have been entered, the polygon is drawn in the picture box when you choose the "Draw" button. Choosing the "OK" button closes the dialog box.

In the Keyboard mode, choose the "Circle" tool to draw a circle in the picture box. A Specify Circle Dimensions dialog box appears prompting for information about the circle center coordinates and radius. Specify the type of circle coordinates you are going to enter by selecting one of the two Coordinate Type options: "Screen" or "Site". Type the circle center coordinates and horizontal radius in the appropriate text boxes. The circle is drawn in the picture box when you choose the "Draw" button. Choosing "Next" clears the text boxes and resets the focus so that additional circles can be drawn. Choosing the "OK" button closes the dialog box.

In the Mouse or Keyboard mode, choose the "Ellipse" tool to draw an ellipse in the picture box. A Specify Ellipse Dimensions dialog box appears prompting for information about the center coordinates, radius, and aspect ratio of the ellipse. Specify the type of ellipse coordinates you are going to enter by selecting one of the two Coordinate Type options: "Screen" or "Site". Type the center coordinates, horizontal radius, and the aspect ratio of the ellipse in the appropriate text boxes. A circle has an aspect ratio of 1. Integer aspect ratio values such as 2, 3, and 4 are typed when you want the Y-axis of the ellipse to be larger than the X-axis. Fractional aspect ratio values such as 0.5, 0.333, and 0.25 are typed when you want the X-axis of the ellipse to be larger than the Y-axis. If the aspect ratio is greater than 1, the ellipse is stretched in the vertical direction. If the aspect ratio is less than 1, the ellipse is stretched in the horizontal direction. As the aspect ratio gets larger, the ellipse gets closer and closer to being a vertical line. The ellipse is drawn in the picture box when you choose the "Draw" button. Choosing "Next" clears the text boxes and resets the focus so that additional ellipses can be drawn. Choosing the "OK" button closes the dialog box.

In the Mouse or Keyboard mode, choose the "Arc" tool to draw an arc in the picture box. A Specify Arc Dimensions dialog box appears prompting for information about the center coordinates, radius, and starting and ending angles of the arc. Specify the type of arc coordinates you are going to enter by selecting one of the two Coordinate Type options: "Screen" or "Site". Type the center coordinates, horizontal radius, start angle (in degrees), and the end angle (in degrees) of the arc in the appropriate text boxes. Angles are measured counterclockwise from the positive X-axis. The arc is drawn in the picture box when you choose the "Draw" button. Choosing "Next" clears the text boxes and resets the focus so that additional arcs can be drawn. Choosing the "OK" button closes the dialog box.

In the Mouse or Keyboard Mode, choose the "Curve" tool to draw a spline curve in the picture box. A Specify Curve Dimensions dialog box appears prompting for information about the coordinates of points on the curve. Specify the type of curve coordinates you are going to enter by selecting one of the two Coordinate Type options: "Screen" or "Site". Type the number of points on the curve and the screen or site point coordinates in the appropriate text boxes. The number of points must be at least 6 and cannot exceed 100. X-coordinates must increase with point numbers so that the X-coordinate for point number 2 is greater than the X-coordinate for point number 1, the X-coordinate for point number 3 is greater than the X-coordinate for point number 2, etc. Curves that do not conform to this rule should be divided into two or more curve segments that do conform to this rule. Choosing the "Enter" button stores coordinates for the point whose number is displayed. Choosing the "Next" button increments the point number, clears text boxes, and resets the focus so that data for additional points can be entered. After you have entered coordinates for all points, the curve is drawn in the picture box when you choose the "Draw" button. Choosing the "OK" button closes the dialog box.

In the Keyboard mode, choose the "Text" tool to add text in the picture box. A Specify Text Dimensions dialog box appears prompting for information about the text string. Specify the type of text coordinates you are going to enter by selecting one of the two Coordinate Type options: "Screen" or "Site". Type the upper left text string corner coordinates and the text string in the appropriate text boxes. The text string is displayed in the picture box when you choose the "Enter" button. Choosing the "Next" button clears text boxes and resets the focus so that additional text can be added to the picture box. Choosing the "OK" button closes the dialog box.

In the Keyboard mode, choose the "Erase" tool to remove objects, entities, and text from a box area in the picture box. A Specify Erase Dimensions dialog box appears prompting for information about the box area definition points. Specify the type of box area coordinates you are going to enter by selecting one of the two Coordinate Type options: "Screen" or "Site". Type the screen or site upper left and lower right corner box area X- and Y-coordinates in the appropriate text boxes. Objects, entities, and text in the box area are erased from the picture box when you choose the "Erase" button. Choosing the "Next" button clears the text boxes and resets the focus so that other objects, entities, and text can be erased. Choosing the "OK" button closes the dialog box.

Cross Section

Choosing the "Log" tool displays a Specify Log Dimensions data entry dialog box that prompts for information about the well cross section to be drawn in the Design Window and displays a well number. Well numbers are synchronized with those entered in the Specify Cross Section Dimensions dialog box. Specify the log type by selecting one of the two Cross Section Type options: "Geologic" or "Hydrogeologic". Specify whether or not logs should be labeled by selecting one of the two Log Display options: "Label" and "Do Not Label". If "Label" is selected, well numbers will be printed at the top of the log and code names will be printed to the right of the log except where the printing of code names will create unreasonable clutter. Type the first unit number associated with the displayed well number in the appropriate text box, and select a color code to represent that unit on the cross section to be drawn from the Color List box using the scroll bars or arrow keys. Any sequence of units consistent with imported or edited log data can be chosen.

A palette of 16 colors — Black, Blue, Green, Cyan, Red, Magenta, Yellow, White, Gray, Light Blue, Light Green, Light Cyan, Light Red, Light Magenta, Light Yellow, and Bright White — is supported by the Specify Log Dimensions dialog box Color List box. A sample of the unit color code is displayed after a color is selected. Choosing the "Enter" button stores the selected well and unit numbers and color. Choosing the "Draw" button displays a line with the selected color in the picture box at the location of the well from the top to the base of the selected unit. Choosing the "Next Unit" button clears the unit number text box and resets the focus so that another unit can be color coded and drawn on the cross section. Choosing the "Next Well" button increments the well number to that corresponding to the next well on the cross section, clears text boxes, and resets the focus so that units in additional wells can be color coded and drawn on the cross section. Choosing the "OK" button closes the dialog box.

After all logs on the cross section have been drawn, hydrostatigraphic and hydrogeologic units can be defined by connecting appropriate geologic unit tops and bases with the Design Window Line tool. The Undo tool can be used to erase undesirable lines. Undo clears the picture box and displays the prior image that appeared in the picture box; in other words, it erases the latest objects, entities, or text placed in the picture box. Other Design Window tools can be used to add text and objects to the cross section, such as water level symbols and elevations.

Choosing the "Model" tool results in the appearance of a Specify Log Dimensions dialog box that requests information about the model cross section to be drawn in the picture box and displays a model number corresponding to one of the model numbers entered in the Specify Cross Section Dimensions dialog box. Specify whether or not the model log should be labeled by selecting one of the two Log Display options: "Label" or "Do Not Label". Type the first layer number associated with the displayed model number in the appropriate text box, and select a color code to represent that layer from the Specify Log Dimensions dialog box Color List box using the scroll bars or arrow keys. Any sequence of layers consistent with the imported or edited log data can be chosen.

A palette of 16 colors — Black, Blue, Green, Cyan, Red, Magenta, Yellow, White, Gray, Light Blue, Light Green, Light Cyan, Light Red, Light Magenta, Light Yellow, and Bright White — is supported by the Color List box. A sample of the unit color code is displayed after a color is selected. Choosing the "Enter" button stores the selected layer number and color. Choosing the "Draw" button displays a thick line with the specified color in the picture box from the top to the base of the chosen layer. Choosing the "Next Layer" button clears the layer number text box and resets the focus so that another layer can be color coded and drawn on the cross section. Choosing the "Next Model" button increments the model number, clears text boxes, and resets the focus so that additional models can be color coded and drawn on the cross section. Choosing the "OK" button closes the dialog box.

Choosing the "WL" tool displays a Water Level Data Display data entry dialog box prompting for information about the well or model and unit or layer for which water level data are to be displayed. Type the desired well or model number and the unit or layer number in the appropriate text boxes. Choosing the "Enter" button displays the water level elevation and unit in the dialog box. The displayed water level elevation can be noted and used to draw and label a water level symbol on the cross section with Design Window object and text tools. Choosing the "Next" button clears text boxes and resets the focus so that water level data for additional wells or models and units or layers can be viewed. Choosing the "OK" button closes the dialog box.

Choosing the "Chem" tool displays a Geochemical Data Display dialog box prompting for information about the well and unit for which concentration data are to be displayed. Type the well, unit, and constituent numbers in the appropriate text boxes. Choosing the "Enter" button displays the constituent number, name, unit, and concentration in the dialog box. The displayed information can be noted and used to draw and label a constituent symbol on the cross section with the Design Window object and text tools. Choosing the "Next" button clears text boxes and resets the focus so that data for another well, unit, and constituent can be viewed. Choosing the "OK" button closes the dialog box.

The Design Window picture box image can be saved in a file by choosing the "Save" tool. A Save Image dialog box is displayed. Type a valid name for the file in which the picture box image is to be saved, such as C:\DM\DMB1.BMP, in the Save As text box. Choosing the "OK" button opens the file, saves the picture box image in the bitmap format, closes the file, and closes the dialog box. Only the *.BMP format is supported by *DesignMod*. Be aware that the length of a Design Window image file can exceed 85,000 bytes.

The Design Window picture box image can be printed using the saved image file and *Windows Paintbrush* accessory (see Chapter 10, "Paintbrush", in Microsoft's *Windows User's Guide*). Start *Paintbrush* (see *Windows Primer*). From the File menu choose the "Open" option. In the File Name text box, type the name of the file in which the picture box image was saved. Choose the "OK" button. From the File menu, choose the "Print" and/or the "Printer Setup" option(s). Make any necessary changes to the options. Choose the "OK" button and the image is printed.

A saved image can be displayed in the Design Window picture box by using the Load tool. Choosing the "Load" tool displays a Specify Filename dialog box. You can examine your file system and select among available files with the Drive List box, Directory List box, Wildcard Filename Extension box, and File List box. Alternatively, you can specify a valid name for the file to be loaded in one of two ways: type the appropriate filename in the Filename text box and choose the "OK" button to store the filename, or use the Drive List box scroll bar arrow to select the desired drive, select the desired directory in the Directory List box, edit the default *.* extension in the Wildcard Filename Extension box using the Extension button (thereby limiting your Filename List box search), use the Filename List box scroll bar arrows to select the desired filename, and choose the "OK" button to store the filename. Choosing the "OK" button opens the file, loads the bitmap image contents of the file onto the picture box, closes the file, and closes the dialog box. Only the *.BMP format is supported by *DesignMod*. The number of wells and the well numbers for which data are stored in the file must coincide with the number of wells and well numbers specified for the Design Window.

Choose the "OK" tool to close the Design Window and return to the Manager.

Plan View

A trial model grid must be drawn on the picture box before the following tools can be used: Info, Cell, Zone, Legend, and Map. The origin (0,0) of the grid will be the upper left corner of the picture box. The outer grid lines will appear along picture box borders, and cells nodes will be at the center of grid line blocks. If the length of the plan view along the X-axis is larger than the length of the plan view along the Y-axis, then the grid will fill the entire picture box width, but the grid will only partially fill the picture box height. If the length of the plan view along the Y-axis is larger than the length of the plan view along the X-axis, then the grid will fill the entire picture box height, but the grid will only partially fill the picture box width. The accumulative column grid spacings must equal the plan view width entered in the Specify Plan View Dimensions dialog box. The accumulative row grid spacings must equal the plan view height entered in the Specify Plan View Dimensions dialog box.

Choosing "Grid" displays a Specify Grid Spacing Dimensions data entry dialog box prompting for information about the trial model grid column and row grid spacings. The column block number 1 appears at the top of the dialog box. Uniform or variable grid spacings and the block method of entering grid spacings are supported. Type the first and last column numbers and the column grid spacing for column block number 1 in the appropriate Block Column Data text boxes. Choosing the "Enter Col" button stores column entries. Choosing the "Next Col" button clears text boxes, resets the focus, and increments the block number so that the grid spacing for another column block can be entered. Type the first and last column numbers and the column grid spacing for column block number 2 in the appropriate text boxes. Choosing the "Enter Col" button stores column entries. Repeat this process until all column block grid spacings have been entered. Choosing the "Row" button changes the block number to 1, indicating that data for row block 1 data can be entered, and resets the focus in the Block Row Data text box. Type the first and last row numbers and the row grid spacing for row block number 1 in the appropriate Block Row Data text boxes. Choosing the "Enter Row" button stores row entries. Choosing the "Next Row" button clears text boxes, resets the focus, and increments the block number so that the grid spacing for another row block can be entered. Type the first and last row numbers and the row grid spacing for row block number 2 in the appropriate text boxes. Choosing the "Enter Row" button stores row entries. Repeat this process until all row block grid spacings have been entered. Type the model grid left column, right column, upper row, and lower row numbers in appropriate text boxes at the bottom of the dialog box. Choosing the "Enter" button stores model grid data. Choosing the "Draw" button draws the trial model grid in the Design Window picture box. Choosing the "OK" button closes the dialog box.

A small color-coded circle can be drawn at points where a particular type of information is available using the Info tool. The description of the information point can be viewed at any time by using the Legend tool. Choosing the "Info" tool displays a Specify Information Point Dimensions data entry dialog box prompting for information about the data point. The type number 1 and the point number 1 are displayed. Specify the type of coordinates that will be entered by selecting one of the two Coordinate Type options: "Screen" or "Site". Type numbers 1–16 and point numbers 1–100 are supported. Select a color to represent the information type from the Color List box using the scroll bars or arrow keys. A palette of 14 colors — Black, Blue, Green, Cyan, Red,

Magenta, Yellow, White, Light Blue, Light Green, Light Cyan, Light Red, Light Magenta, and Light Yellow — is supported. A sample of the selected information type color is displayed. Type the screen or site X- and Y-coordinates for the displayed point number and the information description in the appropriate text boxes. A total of 24 description characters are supported.

Choosing the "Enter" button stores numbers and characters and sets option selection flags. Choosing the "Draw" button displays a small colored circle at the location of the information point on the plan view. Choosing the "Next Point" button increments the point number, clears information point text boxes, and resets the focus so that additional points with the same information type color can be displayed. Choosing the "Next Type" button increments the type number, sets the point number to 1, clears information point text boxes, and resets the focus so that points with additional information type colors can be displayed. Choosing the "OK" button closes the dialog box.

Model cells containing such features as sources, sinks, drains, streams, and boundaries can be highlighted with colors by using the Cell tool. Feature descriptions can be viewed at any time by using the Legend tool. Choosing the "Cell" tool displays a Specify Cell Block Dimensions data entry dialog box prompting for information about the cells being highlighted. The type number 1 and the block number 1 are displayed. A block can represent one or more cells with the same feature. Type numbers 1–14 and block numbers 1–100 are supported. Select a color to represent the feature type from the Color List box using the scroll bars or arrow keys. A palette of 14 colors — Black, Blue, Green, Cyan, Red, Magenta, Yellow, White, Light Blue, Light Green, Light Cyan, Light Red, Light Magenta, and Light Yellow — is supported. A sample of the selected feature type color is displayed. Type the feature description and the upper left and lower right corner column and row numbers of the block in the appropriate text boxes. Column and row numbers 1–200 and 24 description characters are supported.

Choosing the "Enter" button stores numbers and characters and sets option selection flags. Choosing the "Draw" button displays a color block on the plan view representing a particular feature type. Choosing the "Next Block" button increments the block number, clears text boxes, and resets the focus so that additional feature blocks with the same type color can be displayed. Choosing the "Next Type" button increments the type number, sets the block number to 1, clears text boxes, and resets the focus so that features with additional type colors can be displayed. Choosing the "OK" button closes the dialog box.

Model cells within selected Z value zonal blocks can be highlighted with colors using the Zone tool. Z value descriptions such as hydraulic conductivity and Z values can be viewed at any time using the Legend tool. Choosing the "Zone" tool displays a Specify Zone Block Dimensions data entry dialog box prompting for information about the zone being highlighted. The type number 1 and the block number 1 are displayed. Type numbers 1–14 and block numbers 1–100 are supported. Select a color to represent the Z value type from the Color List box using the scroll bars or arrow keys. A palette of 14 colors — Black, Blue, Green, Cyan, Red, Magenta, Yellow, White, Light Blue, Light Green, Light Cyan, Light Red, Light Magenta, and Light Yellow — is supported. A sample of the selected Z value zonal block color code is displayed. Type the Z value description, the upper left and lower right corner column and row numbers of the block, and the Z value in the appropriate text boxes. Column and row numbers 1–200 and 24 description characters are supported.

Choosing the "Enter" button stores numbers and characters and sets option selection flags. Choosing the "Draw" button displays a color block on the plan view representing a particular Z value zone. Choosing the "Next Block" button increments the block number, clears text boxes, and resets the focus so that additional Z value zonal blocks with the same type color can be displayed. Choosing the "Next Type" button increments the type number, sets the block number to 1, clears text boxes, and resets the focus so that Z value zonal blocks with additional type colors can be displayed. Choosing the "OK" button closes the dialog box.

Zone, cell, or information point descriptions and Z values can be viewed by using the Legend tool. Choosing the "Legend" tool displays a Legend Option dialog box offering five Type options — Zone, Cell, Info, Contour, and Zonal Diagram — and two Switch options, On or Off. After choosing one of the "Type" options and the "On Switch" option, move the mouse cursor to a zone, cell, or point of interest and depress the left mouse button. The zone description and Z value or the cell or point description are displayed below the picture box in the Plan View Design Window. Choose the "Legend" tool, the appropriate type option, and the "Off Switch" to turn off the legend display when the mouse button is depressed.

Surface and isopach maps can be superposed on the plan view to provide model design background information by using the Map tool. Choosing the "Map" tool displays a Map menu containing a Title Bar with the window title, a Menu Bar listing available menus, a Control Menu box, and a mouse pointer. The Map Menu Bar menus are File, Data, Interpret, Overlay, and Help. Choosing any one of these menus displays a drop-down menu with commands or options, a dialog box, or a help message. Choosing a drop-down menu item displays a dialog box or a cascading menu containing options. Selecting a cascading menu item displays a dialog box or another cascading menu containing options. Selecting one of these options displays a dialog box.

Choosing "File" displays a cascading menu with four commands: Load, for importing plan view files; Save, for exporting plan view files; Exit, for closing the Map menu; and Help, for viewing information about the other three commands. Choosing "Load" or "Save" eventually displays a dialog box. Choosing the "Cancel" button in any dialog box ends the dialog before completion and closes the dialog box. Choosing the "Help" button in a dialog box displays a Help message box with information concerning the dialog box. Choosing the Help message box "OK" button closes the message box.

Choosing "File" and "Load" displays a cascading menu with two options: Gridded Triplet and Scattered Triplet. Gridded Triplet refers to an array of XYZ values with upper left origin and model coordinates covering the plan view grid, and Scattered Triplet refers to an array of XYZ values for points within the plan view grid with lower left origin and site base map coordinates. Choosing "File", "Load", and "Gridded Triplet" displays a Specify Filename dialog box. Type a valid name such as C:\DM\DM1.DAT for the plan view file to be imported in the Filename text box. Choosing the "OK" button opens the file, reads the file, closes the file, closes the dialog box, and returns to the Map menu. Choosing "File", "Load", and "Scattered Triplet" displays a Specify Well Or Point Dimensions data entry dialog box prompting for information about the scattered wells and/or points for which data are stored in the file to be imported. Specify the type of data in the file to be imported by selecting one of the three Data Type options: Wells, Points, and Both. Type in the number of wells and/or number of points and a valid name of the file to be imported in the appropriate text boxes. The number of wells and points combined cannot exceed 100. Choosing the "Enter" button stores entries. Choosing the "OK" button opens a file, reads the file, closes the file, closes the dialog box, and returns to the Map menu.

All or any portion of a plan view array stored in the computer's memory can be saved in a file. Choosing "File" from the Map menu and then the drop-down Save menu displays a cascading menu with three options: Gridded Triplet, Scattered Triplet, and Grid. Gridded Triplet refers to an array of XYZ values covering the plan view grid with upper left origin and model coordinates, and Scattered Triplet refers to an array of XYZ values for wells and/or points within the plan view grid with lower left origin and site base map coordinates. Grid refers to an array of Z values covering the plan view grid.

Choosing "File", "Save", and "Gridded Triplet" displays a Specify Filename dialog box requesting the name of the file in which the array is to be saved. Type a valid filename such as C:\DM\DM1G.DAT in the Save As text box. Choosing the "OK" button opens the file, writes the array, closes the file, closes the dialog box, and returns to the Map menu.

Selecting "File", "Save", and "Scattered Triplet" displays a Specify Wells Or Point Dimensions data entry dialog box prompting for information about the scattered wells and/or points for which data are stored in the file to be saved. Specify the type of data in the file to be saved by selecting one of the three Data Type options: Wells, Points, and Both. Type in the number of wells and/or number of points and a valid name such as C:\DM\DM1H.DAT for the file to be saved as in the appropriate text boxes. The number of wells and points combined cannot exceed 100. Choosing the "Enter" button stores entries. Choosing the "OK" button opens a file, writes the array, closes the file, closes the dialog box, and returns to the Map menu.

Selecting "File", "Save", and "Grid" displays a cascading menu with three options: CSV, LSSV, and Binary. CSV refers to comma-separated value file format, LSSV refers to listed space-separated value file format, and Binary refers to binary file format. Selecting any one of these options displays a Specify File Dimensions dialog box requesting the name of the file in which the array is to be saved. Type a valid filename such as C:\DM\DM1G.DAT in the Save As text box. Choosing the "OK" button opens the file, writes the array, closes the file, closes the dialog box, and returns to the Map menu.

A well and/or point database must be created or imported with either *DesignMod*'s Manager File or Edit menu prior to choosing Data, which enables you to identify wells and/or points whose data are to be used for subsequent surface and isopach interpolation. Choosing "Data" displays a drop-down menu with three options: Surface, Isopach, and Help. Choosing "Data" and "Surface" displays a cascading menu with three options: Well, Point, and Well And Point. Choosing "Data", "Surface", and "Well" displays a Specify Surface Well Dimensions data entry dialog box prompting for information about the wells to be used in surface interpolation. Specify the type of logs to be used in surface interpolation by selecting one of the three Log Type options: "Geologic", "Hydrogeologic", or "Geochemical". Specify the type of surface to be interpolated by selecting one of the four Surface Type options: "Top", "Base", "Water Level", or "Concentration". Type the number of wells to be used in surface interpolation, the first well number, and the first well unit number in appropriate text boxes. Well numbers 1–100 and unit numbers 1–25 are supported if the Geologic or Hydrogeologic Log Type was selected. Well numbers 1–100, unit numbers 1–10, and constituent numbers 1–25 are supported if the Geochemical Log Type was selected. Choosing the "Enter" button stores entries. Choosing the "Next" button clears text boxes and resets the focus so that additional well and unit numbers can be entered. Unit numbers in all wells must be associated with the same surface. Choosing the "OK" button closes the dialog box and returns to the Map menu.

Choosing "Data", "Surface", and "Point" displays a Specify Surface Scattered Point Dimensions data entry dialog box prompting for information about the surface points to be used in interpolation. The point number is displayed. Type the X- and Y-coordinates of the displayed point number and the point Z value in appropriate text boxes. Point numbers 1–500 are supported. Choosing the "Enter" button stores entries. Choosing the "Next" button clears text boxes and resets the focus so that data for additional points can be entered. Choosing the "OK" button closes the dialog box and returns to the Map menu.

Choosing "Data", "Surface", and "Well and Point" first displays a Specify Surface Well Dimensions data entry dialog box and later a Specify Surface Scattered Points Dimensions data entry dialog box so that both well and point data can be entered for subsequent surface interpolation as discussed earlier. The number of wells and points combined cannot exceed 100.

Choosing "Data" and "Isopach" displays a Specify Isopach Dimensions data entry dialog box prompting for information about the unit for which the thickness is to be interpolated. Type the number of wells to be used in the isopach interpolations in the appropriate text box. Well numbers 1–100 are supported. Specify the type of log to be used in interpolation by selecting one of the two Log Type options: "Geologic" or "Hydrogeologic". Type the first well number, unit number, and a valid name such as C:\DM\DM1T.DAT for the file in which interpolated isopach values are to be stored in the appropriate text boxes. Choosing the "Enter" button stores these entries. Choosing the "Next" button clears text boxes and resets the focus so that data for additional wells can be entered. Choosing the "OK" button calculates and stores interpolated thicknesses at wells, opens the file, writes thicknesses (Z values), closes the file, closes the dialog box, and returns to the Map menu.

Scattered or gridded well, point, or isopach data must be created with the Data menu or imported with the File menu before using the Interpret menu to generate a regular (uniform or variable) grid array of surface Z values. Triangulation and kriging interpretation techniques are supported. The generated grid Z value array covering the plan view grid can be superposed over the plan view map and used as background information. Choosing the "Interpret" menu displays a Specify Interpretation Dimensions dialog box offering interpretation technique options. Specify the type of interpretation techniques that will be used by selecting one of the two Interpretation Technique options: Triangulation or Kriging. If "Kriging" is selected, specify the type of kriging results that will be generated by selecting one of the two Kriging Results options: Z Values or Error. Choosing the "Enter" button stores selections and calculates Z values at grid nodes using the triangulation or kriging technique. Be aware that the calculation time will be long if the grid contains a large number of columns and rows. Choosing the "OK" button closes the dialog box and returns to the Map menu.

Before selecting the Overlay menu, a regular gridded triplet array must be created with the Interpret menu or imported with the File menu. Selecting the "Overlay" menu displays a drop-down menu with three options: Contour, Zone, and Help. Contour refers to drawing surface Z or isopach values on the plan view or on paper. Zone refers to drawing zonation cell color patterns on the plan view or typing letter zonation cell codes on paper.

Choosing "Overlay" and "Contour" displays a cascading menu with two options: Screen and Printer. Screen refers to the drawing of contours on the plan view, and Printer refers to printing contours on papers. Selecting "Overlay", "Contour", and "Screen" displays a Specify Contour Map Dimensions data entry dialog box prompting for information about contour values. The contour number is displayed. Type the number of contours to be drawn on the plan view and the Z or isopach value of the contour whose number is displayed in the appropriate text boxes. Contours numbers 1–12 are supported. Choosing the "Enter" button stores these entries. Choosing the "Next" button clears text boxes, increments the contour number, and resets the focus so that additional contour values can be entered. Choosing the "Plot" button draws Z value contours with colors on the plan view. Be aware the calculation time will be long if the grid has a large number of columns and rows. Choosing the "OK" button closes the dialog box and returns to the Map menu.

Contour values can be viewed by using the Legend tool. Choosing the "Legend" tool displays a Legend Option dialog box offering five Type options — Zone, Cell, Info, Contour, and Zonal Diagram — and two Switch options, On and Off. Choose the "Contour Type" option and the "On Switch" option. Position the mouse cursor over a contour and depress the left mouse button. The value of the contour is displayed below the picture box in the Plan View Design Window.

Choosing "Overlay", "Contour", and "Printer" displays a Specify Printer Contour Map Dimensions data entry dialog box prompting for information about contour values. The contour number is displayed. Type the X- and Y-coordinate unit, the Z value unit, and the value of the contour whose number is displayed in the appropriate text boxes. Contour numbers 1–12 and 12 unit characters are supported. Choosing the "Unit" button stores unit characters. Choosing the "Contour" button stores the contour value. Choosing the "Next" button clears text boxes, increments the contour number, and resets the focus so that additional contour values can be entered. Choosing the "Print" button draws the contour map on paper. Be aware that calculation time will be long if the grid contains a large number of columns and rows. Choosing the "OK" button closes the dialog box and returns to the Map menu.

Selecting "Overlay", "Zone", and "Screen" displays a Specify Zonal Diagram Screen Dimensions data entry dialog box prompting for information about zonation ranges. The range number is displayed. Type the number of zonation ranges and the upper and lower range values for the displayed range number in appropriate text boxes. Range numbers 1–12 are supported. Choosing the "Enter" button stores entries. Choosing the "Next" button clears text boxes, increments the range number, and resets the focus so that additional range values can be entered. Choosing the "Plot" button draws color-coded cell blocks representing zonation ranges on the plan view. Choosing the "OK" button closes the dialog box and returns to the Map menu.

Z value range values can be viewed by using the Legend tool. Choosing the "Legend" tool displays a Legend Option dialog box offering five Type options — Zone, Cell, Info, Contour, and Zonal Diagram — and two Switch options, On and Off. Choose the "Zonal Diagram" Type option and the "On Switch" option. Position the mouse cursor over a cell and depress the left mouse button. The Z value range for the cell is displayed below the picture box in the Plan View Design Window.

Selecting "Overlay", "Zone", and "Printer" displays a Specify Printer Zonal Diagram Dimensions data entry dialog box prompting for information about range values. The Z value range number is displayed. Type the Z value unit and upper and lower values for the displayed range in the appropriate text boxes. Range numbers 1–12 and 12 unit characters are supported. Choosing the "Unit" button stores unit characters. Choosing the "Range" button stores range values. Choosing the "Next Range" button clears text boxes, increments the range number, and resets the focus so that additional range values can be entered. Choosing the "Print" button types zonation coded letters on paper. Choosing the "OK" button closes the dialog box and returns to the Map menu.

The Plan View Design Window picture box image can be saved in a file by choosing the "Save" tool. A Save Image dialog box is displayed. Type a valid name such as C:\DM\DM1B.BMP for the file in which the image is to be saved in the Save As text box. Choosing the "OK" button opens the file, saves the image in the bitmap format, closes the file, and closes the dialog box. Only the *.BMP format is supported by *DesignMod*. Be aware that the size of a Plan View Design Window image file can exceed 85,000 bytes.

A Plan View Design Window picture box image can be printed using the saved image file and *Windows Paintbrush* accessory (see Chapter 10, "Paintbrush", in Microsoft's *Windows User's Guide*). Start *Paintbrush* (see *Windows Primer*). From the File menu, choose the "Open" option. In the File Name text box, type the name of the file in which the picture box image was saved. Choose the "OK" button. From the File menu, choose the "Print" and/or the "Printer Setup" option(s). Make any necessary changes to the options. Choose the "OK" button and the image is printed.

A saved image can be displayed in the Plan View Design Window picture box by using the Load tool. Choosing the "Load" tool displays a Specify Filename dialog box. You can examine your file system and select among available files with the Drive List box, Directory List box, Wildcard Filename Extension box, and File List box. Alternatively, you can specify a valid name such as C:\DM\DM1B.BMP for the file that is to be loaded in one of two ways: (1) type the appropriate filename in the Filename text box and choose the "OK" button to store the filename, or (2) use the Drive List box scroll bar arrow to select the desired drive, select the desired directory in the Directory List box, edit the default *.* extension in the Wildcard Filename Extension box using the Extension button, thereby limiting your Filename List box search, use the Filename List box scroll bar arrows to select the desired filename, and choose the "OK" button to store the filename. Choosing the "OK" button opens the file, loads the bitmap image contents of the file onto the plan view picture box, closes the file, and closes the dialog box. Only the *.BMP format is supported by *DesignMod*. The plan view dimensions incorporated into the data that are stored in the file must coincide with the plan view dimensions specified for the Design Window.

Choose the "OK" tool to close the Design Window and return to the Manager.

EXERCISES

The following exercises are designed for easy and rapid completion and sampling of the main operating features of *DesignMod*. Only simple cross

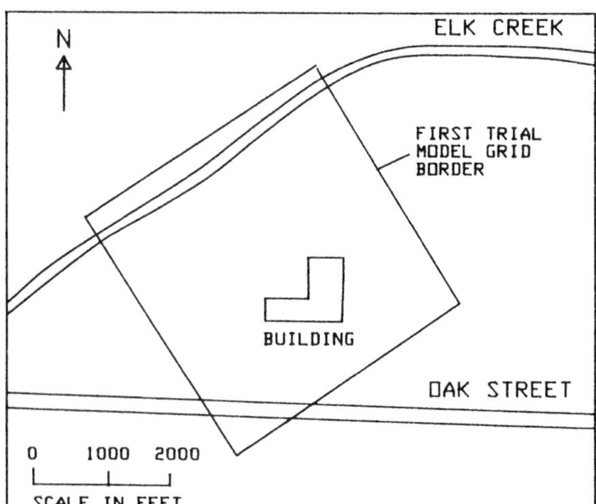

Figure 1.   Area of interest for Exercise DM1.

sections, plan views, and arrays are included in the exercise scenarios. After practicing with these exercises, you should have little trouble addressing your own complex scenarios. (*Note:* Do not type quote marks in the text boxes.)

DM1 — Cross Section

Exercise DM1 helps you master the use of *DesignMod*'s File, Edit, and Draw menus and associated data entry dialog boxes and design window. You will create and save a well log database; draw geologic, hydrogeologic, and model cross sections based on the database; and define the location and orientation of a first trial model grid for use in Exercise DM2. The area of interest for Exercise DM1 (Figure 1), previously defined on a paper site base map, is underlain by about 70 feet of unconsolidated glaciofluvial sediments which overlie shale bedrock. The geologic logs for five wells along a cross section (see Figure 5) through the area of interest are given in Table 3.

Water table elevations at well numbers 24, 17, 14, 6, and 2 are 557.12, 559.53, 560.28, 561.71, and 562.23 feet, respectively. Water level elevations in the sand and gravel deposits immediately above bedrock at well numbers 24, 17, 14, 6, and 2 are 555.38, 556.55, 555.62, 552.06, and 552.44 feet, respectively. The length of the cross section is about 0.5 mile. Site base map well coordinates are given in Table 4.

If you are in a hurry, you may wish to import a file containing this well log database in the *DesignMod* database format using the File menu instead of entering the well log database using the Edit menu. If so, start *DesignMod* and choose the "File" menu in the Manager Menu Bar. Select the "Load" drop-down menu option and the "Log" and "Geology" cascading menu options. A

Table 3. Well Log Data for Exercise DM1

Material	Top Elevation (ft)	Base Elevation (ft)	Thickness (ft)
Well Number 24			
Fill	564	556	8
Sand	556	538	18
Clayey sand	538	535	3
Sand	535	525	10
Clay and clayey silt	525	512	13
Sand	512	511	1
Clay and clayey silt	511	497	14
Sand and gravel	497	495	2
Shale	495	400	95
Well Number 17			
Silty sand	565	559	6
Clayey sand	559	557	2
Silty sand	557	553	4
Sand	553	521	32
Silt	521	520	1
Well Number 14			
Silty sand	566	559	7
Sand	559	549	10
Silt	549	547	2
Sand	547	529	18
Clayey sand	529	528	1
Sand	528	525	3
Silt and sandy silt	525	507	18
Sand	507	505	2
Silt and sandy silt	505	499	6
Sand and gravel	499	494	5
Shale	494	398	96
Well Number 6			
Fill	569	565	4
Silty sand	565	556	9
Sand	556	548	8
Silt	548	546	2
Sand	546	541	5
Silt and sandy clay	541	516	25
Sand and gravel	516	494	22
Shale	494	395	99

Table 3 (continued). Well Log Data for Exercise DM1

Material	Top Elevation (ft)	Base Elevation (ft)	Thickness (ft)
Well Number 2			
Silty sand	570	558	12
Sand	558	545	13
Silty sand and clay	545	532	13
Silt and sandy silt	532	515	17
Sand	515	508	7
Sand and gravel	508	492	16
Shale	492	392	100

Table 4. Well Coordinate Data for Exercise DM1

Well Number	X-Coordinate (ft)	Y-Coordinate (ft)
24	3002	4397
17	3685	3772
14	4023	3505
6	4378	2998
2	5044	2856

Specify Filename dialog box appears, prompting for the name of the file in which the well log database is stored. The well log database file for Exercise DM1 is stored in file DM1G.DAT on the disk. Type C:\DM\DM1G.DAT or a similar specification in the Filename text box, or choose the appropriate drive from the Drive List box and the appropriate directory from the Directory List box. Edit the Wildcard File List Extension text box to read *.DAT to limit the filenames displayed in the File List box. Choose the filename DM1G.DAT from the File List box.

After typing or choosing a filename, choose the "OK" button to close the dialog box and display a Specify File Dimensions data entry dialog box prompting for information about the well and unit numbers for which well log data are stored in file DM1G.DAT. Type "5" in the Number Of Wells Or Models text box, "24" in the Well Or Model Number text box, and "9" in the Number Of Units Or Layers text box. Choose the "Enter Well Or Model" button to store these entries. Type "1" in the Unit Or Layer Number text box. Choose the "Enter Unit Or Layer" button to store this entry. Choose the "Next Unit Or Layer" button to clear and reset the focus on the Unit Or Layer Number text box. Type "2" in the Unit Or Layer Number text box. Choose the "Enter Unit Or Layer" button to store this entry. Choose the "Next Unit Or Layer" button to clear and reset the focus on the Unit Or Layer Number text box. Repeat this process until all nine unit numbers have been entered. Choose the "Next Well Or Model" button to clear text boxes and reset the focus on the Well Or Model Number text box. Enter data for well numbers 17, 14, 6, and 2 in the manner previously described. Choose the "OK" button to close the dialog box and return to the Manager.

To enter the well log database using the Edit menu, start *DesignMod* and choose the "Edit" menu in the Manager Menu Bar. Select the "Log" drop-down menu option and the "Geology" cascading menu option. A Specify Geology Dimensions data entry dialog box will appear prompting for information about geologic well logs. To enter data for well number 24, type "24" in the Well Number text box, "9" in the Number Of Units text box, "3002" in the X-Coordinate text box, "4397" in the Y-Coordinate text box, and "Feet" in the X and Y Unit text box. Review and revise these entries if necessary and then choose the "Enter Well" button to store these entries. To enter data for well number 24 unit numbers 1–9, type "1" in the Unit Number text box, "Fill" in the Name text box, "FIL" in the Code text box, "564" in the Top Elevation text box, "556" in the Base Elevation text box, and "Feet" in the Elevation Unit text box. Choose the "Enter Unit" button to store these entries. Choose the "Next Unit" button to clear unit text boxes and reset the focus on the Unit Number text box. Enter and store data for well number 24 unit number 2 in the same manner as described earlier. Repeat this process until data for all nine units of well number 24 have been entered. Choose the "Next Well" button to clear text boxes and reset the focus on the Well Number text box so that data for well number 17 can be entered. Repeat the process used in entering data for well number 24 to enter data for well number 17. After entering well log data for the five wells, choose the "OK" button to close the dialog box and return to the Manager.

To save the well log data in a file, choose the "File" menu in the Manager Menu Bar. Select the "Save" drop-down option and the "Log" and "Geology" cascading menu options. A Specify File Dimensions data entry dialog box appears prompting for information about the number of wells and units to be stored in the file. To save data for all five wells and all units in the wells, type "5" in the Number Of Wells Or Models text box, "24" in the Well Or Model Number text box, "9" in the Number Of Units Or Layers text box, and a valid filename such as C:\DM\DM1G.DAT in the Save As text box. Choose the "Enter Well Or Model" button to store these entries.

Type "1" in the Unit Or Layer text box and choose the "Enter Unit Or Layer" button to store this entry. Choose the "Next Unit Or Layer" button to clear the Unit Or Layer Number text box and to reset the focus on the Unit Or Layer Number text box. Type "2" in the Unit Or Layer Number text box and choose the "Enter Unit Or Layer" button to store this entry. Repeat this process until data for all units of well number 24 are entered. Choose the "Next Well Or Model" button to clear text boxes and reset the focus on the well Or Model

Number text box so that data for well number 17 can be entered. Repeat the process used in entering data for well number 24 to enter data for well number 17. After entering well log data for the five wells, choose the "OK" button to open the file, write well log data, close the file, close the dialog box, and return to the Manager.

To view the well log data one well (block) at a time in a spreadsheet box on the screen, choose the "View" menu in the Manager Menu Bar. Select the Tabular drop-down menu option and the "Log", "Geology", and "Screen" cascading menu options. A Specify Tabular Display Dimensions data entry dialog box appears prompting for information about the well number and unit numbers for which log data are to be displayed. To display data for all nine units in well number 24 and all five units in well number 17, type "1" in the Block Number text box, "24" in the Well Or Model Number text box, "1" in the Unit or Layer First Number text box, and "9" in the Unit Or Layer Last Number text box. You could have typed any other combination of numbers 1–9, such as 2 and 7, in the Unit Or Layer First and Last Number text boxes. Choose the "Enter" button to store these entries. Choose the "Next" button to clear text boxes and reset the focus. Type "2" in the Block Number text box, "17" in the Well Or Model Number text box, "1" in the Unit Or Layer First Number text box, and "5" in the Unit Or Layer Last Number text box. Choose the "Enter" button to store these entries and the "OK" button to close the dialog box.

A Tabular Display data entry dialog and spreadsheet box appears prompting for information about the block number and the type of display format. Type "1" in the Block Number text box and select the "Decimal Format Type" option. Choose the "View" button to display the log data for all nine units in well number 24 in a spreadsheet table. Use the scroll bars or arrow keys to view all data. Choose the "Next" button to clear spreadsheet cells and the text box. Type "2" in the Block Number text box. Choose the "View" button to display the log data for all five units in well number 17 in a spreadsheet table. Use the scroll bars or arrow keys to view all data. Choose the "OK" button to close the dialog and spreadsheet box and return to the Manager.

To obtain hard copies of the geologic logs for well numbers 24 and 17, choose the "View" menu in the Manager Menu Bar. Select the "Tabular" drop-down menu option and the "Log", "Geology", and "Printer" cascading menu options. A Specify Log Printer Dimensions data entry dialog box appears prompting for information about the well and unit numbers for which log data are to be printed. Type "24" in the Well Or Model Number text box, "1" in the Unit Or Layer First Number text box, and "9" in the Unit Or Layer Last Number text box. Select the "Decimal Format Type" option button. Type "Feet" in the Unit or Layer X- and Y-Coordinate and Elevation text boxes. Choose the "Print" button to print the log for well number 24. Choose the "Next" button to clear text boxes, reset the focus, and print the log for well number 17. Type "17" in the Well Or Model Number text box, "1" in the Unit or Layer First Number text box, and "5" in the Unit or Layer Last Number text box. Choose the "Print" button to print the log for well number 17. Choose the "OK" button to close the dialog box and return to the Manager.

To draw a cross section based on the previously entered log database, use the following color legend to represent well log materials: Sand and Gravel—Light Blue; Sand—Light Cyan; Silty Sand—Cyan; Silt and Sandy Silt—Light Green; Silt—Green; Silty and Sandy Clay—Light Yellow; Silty Sand and Clay—Yellow; Clay and Clayey Silt—Light Magenta; Clayey Sand—Magenta; Shale—Red; and Fill—Black. To draw well logs on the cross section, choose the "Draw" menu in the Manager Menu Bar. Choose the "Cross Section" drop-down menu option. A Specify Cross Section View Dimensions data entry dialog box appears prompting for information about the upper and lower elevations of the cross section, the wells along the cross section, and the vertical and horizontal units. The block number 1 appears. Choose the "Well Cross Section Type" option. To include all five wells and units except the shale unit in the cross section, type "580" in the Vertical Upper Dimension text box, "480" in the Vertical Lower Dimension text box, "5" in the Number Of Wells Along Cross Section text box, and "24" (the first well along the cross section) in the Well Number text box. Choose the "Enter Well" button to store these entries. You could have included only two, three, or four wells and a limited number of units in the cross section by entering appropriate data in the text boxes. Choose the "Next Well" button to clear and reset the focus on the Well Number text box. Type "17" in the Well Number text box. Choose "Enter Well" to store this entry. Repeat this process until numbers for all wells along the cross section have been entered. Enter "FEET" in the Vertical and Horizontal Unit text boxes. Choose the "Enter Block" button to store these entries. You could choose the "Next" button to enter data for another cross section block containing different wells and/or units. Choose the "OK" button to close the dialog box.

A Cross Section Design Window appears, prompting for the block number. Type "1" in the top center Block Number text box and choose the adjacent "Enter" button to store this entry. Choose the "Log" tool to draw the well logs on the cross section. A Specify Log Dimensions data entry dialog box appears and prompts for information about the wells for which logs are to be drawn on the cross section and displays 24, the number of the first well along the cross section. Select the "Geologic Log Type" option button and the "Label Log Display" option button. To draw all units, type "1" (first unit fill) in the Unit Number text box and select the black color to represent fill. Choose the "Enter" button to store these entries. Choose the "Draw" button to draw the unit on the cross section. Choose the "Next" button to clear and reset the focus on the Unit Number text box. Type "2" in the Unit Number text box and select the light blue color to represent sand. Choose the "Enter" button to stores these entries. Choose the "Draw" button to draw the unit on the cross section. Choose the "Next" button to clear and reset the focus on the Unit Number text box. Repeat

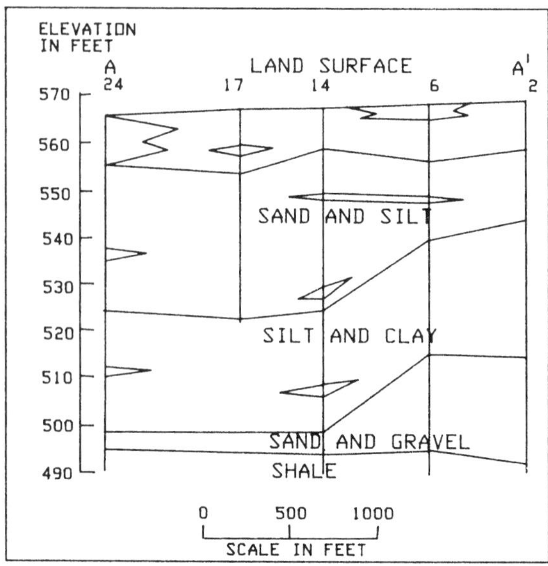

Figure 2.   Cross section for Exercise DM1.

this process until units 1–8 in well number 24 have been drawn (unit 9 is shale and is omitted from the cross section). Choose the "Next Well" button to clear text boxes and reset the focus on the Well Number text box. Enter data for the rest of the wells along the cross section. Choose the "OK" button to close the dialog box and return to the Cross Section Design Window.

Choose the "Key" tool (bottom right tool). A Select Drawing Mode dialog box appears. Select the "Mouse" option and the "OK" button to close the dialog box and return to the Cross Section Design Window. Accept the default line width (1 pixel). Choose the "Poly" tool to draw lines defining each unit that pinches out between wells (lenses) as shown in Figure 2 (see pages 87 to 88). Choose the "Line" tool to draw lines connecting unit tops and bases between wells as shown in Figure 2.

Choose the "Save" tool to save the completed cross section image in a file for future use. A Save data entry dialog box appears prompting for information about the file to be saved. Type a valid filename such as C:\DM\DM1I.BMP in the Save As text box. Choose the "OK" button to save the cross section image and return to the Design Window.

Choose the "Control Menu" box in the upper left corner of the Design Window Title Bar to obtain a hard copy of the cross section using the *Windows Paintbrush* accessory. When the menu appears, choose the "Switch To" drop-down menu option. A Task List dialog box appears. Choose the "Program Manager" in the list box and choose the "Switch To" button to switch from *DesignMod* to Program Manager. Choose the "Accessories Group" icon. An

Accessories Group Window appears. Choose the "Paintbrush Program" icon. The *Paintbrush* window appears. Choose the "File" menu and the "Open" drop-down menu option. An Open dialog box appears. Type the name of the file, such as C:\DM\DM1I.BMP, in which the cross section image was saved in the Filename text box. Choose the "OK" button to close the dialog box. The cross section appears within the Paintbrush Drawing Area. If necessary, use the scroll bars or arrow keys to center the cross section within the *Paintbrush* Drawing Area. Choose the "File" menu, the "Print Setup" menu option (if necessary), and a "Print" menu option. A Print dialog box appears. Choose the "Proof Quality" option and the "Whole Window" option. Accept the default number of copies (1) and scaling (100%). Do not select the Use Printer Resolution option. Choose the "OK" button to obtain a hard copy of the cross section image and close the dialog box.

Choose the "File" menu and the "Exit" drop-down option to close the Paintbrush window. Choose the "Minimize" button in the upper right corner of the Accessories Group Title Bar. Choose the "Control Menu" box in the upper left corner of the Program Manager Title Bar. When the menu appears, choose the "Switch To" drop-down menu option. A Task List dialog box appears. Choose "DesignMod" in the list box and choose the "Switch To" button to switch from Program Manager back to *DesignMod* and the Design Window.

Be aware that you can load a cross section image onto an empty Design Window by choosing the "Load" tool. A Specify Filename dialog box appears. Type the appropriate specification in the Filename text box, and choose the "OK" button to close the dialog box and load the cross section image onto the Design Window. The dimensions of the image and the cross section Design Window must coincide.

Based on a careful study of the geologic cross section in the Design Window, three hydrogeologic units (Shallow Aquifer, Confining Bed, and Deep Aquifer) can be identified and defined as shown by the hydrogeologic well logs in Table 5. The lower portion of the Shallow Aquifer is composed of clean sand of fine to coarse texture with some gravel deposits. The upper portion is composed of dune sand with some silt. The Confining Bed is composed of clayey materials. The Deep Aquifer is composed of outwash sand and gravel deposits. The shale has a very low hydraulic conductivity and can be considered to be an aquiclude.

In Table 5, 999.99 signifies no available water level data. You can enter hydrogeologic log data by using the Edit menu in a manner similar to that used in entering geologic log data. The file DM1H.DAT on the disk contains the hydrogeologic log data which can be imported. You can view the hydrogeologic log data on a cross section in a manner similar to that used in viewing geologic log data.

You can view water level by choosing the "WL" tool in the Hydrogeologic Cross Section Design Window. A Water Level Display dialog box appears. Type the number of the well and the number of the unit for which water level

Table 5. Hydrogeologic Data for Exercise DM1

Hydrogeologic Unit	Top Elevation (ft)	Base Elevation (ft)	Thickness (ft)	Water Level Elevation (ft)
Well Number 24				
Shallow Aquifer	557	525	32	557
Confining Bed	525	498	27	999.99
Deep Aquifer	498	495	3	555
Well Number 17				
Shallow Aquifer	560	521	39	560
Confining Bed	521			
Well Number 14				
Shallow Aquifer	560	525	35	560
Confining Bed	525	499	26	999.99
Deep Aquifer	499	494	5	554
Well Number 6				
Shallow Aquifer	562	541	21	562
Confining Bed	541	516	30	999.99
Deep Aquifer	516	494	23	552
Well Number 2				
Shallow Aquifer	562	545	17	562
Confining Bed	545	515	30	999.99
Deep Aquifer	515	492	23	552

data are to be displayed in appropriate text boxes. Choose the "Enter" button to store these entries and display the water level elevation and the elevation unit. Choose the "Next" button to clear text boxes and reset the focus so that water level data for another well and/or unit can be displayed. Choose the "OK" button to close the dialog box and return to the Cross Section Design Window.

Hydrogeologic logs suggest that the model for the area of interest should have three layers (Aquifer 1, Confining Bed, and Aquifer 2) as defined in Table 6.

In Table 6, 999.99 signifies no available water level data. You can enter model layer data by using the Edit menu in a manner similar to that used in entering geologic log data. The file DM1M.DAT on the disk contains the model layer data, which can be imported. You can view the model layer data on a cross section in a manner similar to that used in viewing geologic log data.

Table 6. Model Layers for Exercise DM1

Layer	Top Elevation (ft)	Base Elevation (ft)	Average Thickness (ft)	Water Level Elevation (ft)
Aquifer 1	560	530	30	560
Confining Bed	530	505	25	999.99
Aquifer 2	505	494	11	555

You can view water level data by choosing the "WL" tool in the Model Cross Section Design Window. A Water Level Display dialog box appears. Type the number of the model and the number of the layer for which water level data are to be displayed in appropriate text boxes. Choose the "Enter" button to store these entries and display the water level elevation and the elevation unit. Choose the "Next" button to clear text boxes and reset the focus so that water level data for another model and/or layer can be displayed. Choose the "OK" button to close the dialog box and return to the Cross Section Design Window.

The Shallow Aquifer is contaminated with two organic compounds, chloroethane and vinyl chloride, migrating from a source located 100 feet northwest of well number 6. A plume of contaminated groundwater has migrated about 250 feet northwest of the source towards nearby private wells. The chloroethane plume is shown in Figure 3. Concentrations of these two contaminants in wells are given in Table 7.

You can enter contaminant concentration data by using the Edit menu in a manner similar to that used in entering hydrogeologic log data. The file DM1C.DAT on the disk contains the contaminant concentration data, which can be imported. You can view the contaminant concentration data by choosing the "Chem" tool with the Hydrogeologic Cross Section Design Window. A Geochemical Data Display dialog box appears prompting for information about the well, unit, and constituent for which data are to be displayed. Type the well number, hydrogeologic unit, and constituent number for which data are to be displayed in the appropriate text boxes. Choose the "Enter" button to display the corresponding constituent name, concentration, and concentration unit data. Choose the "Next" button to view data for additional wells, units, or constituents, or choose the "OK" button to close the dialog box and return to the Hydrogeologic Cross Section Design Window.

In practice, several cross sections at various directions across the area of interest would be drawn and studied in part to determine the location and orientation (upper left and lower right corner coordinates, X-offset, Y-offset, and rotation angle) of a first trial model grid. To save time in completing Exercise DM1, other cross sections will not be drawn. Assume that cross section studies, including that described above, indicate that a first trial model grid should have the following upper left and lower right coordinates, X-offset, Y-offset, and rotation angle: 1100, 4500; 6700, 3200; 3300; 1000; and 32

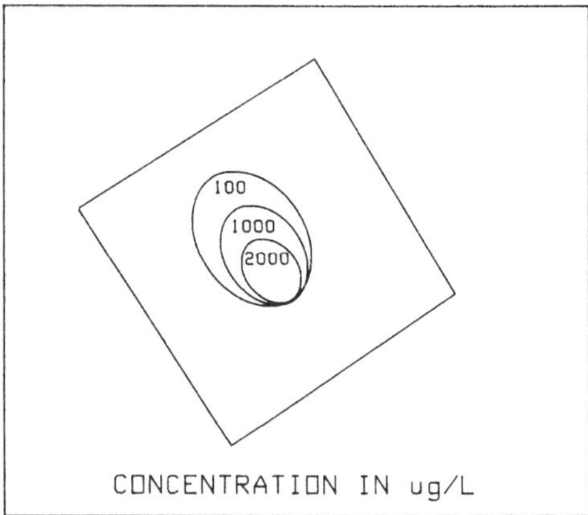

Figure 3.   Chloroethane plume for Exercise DM1.

Table 7. Contaminant Concentration Data for Exercise DM1

Well No.	Chloroethane (micrograms per liter)	Vinyl Chloride (micrograms per liter)
2	0	0
6	0	0
14	3022	345
17	2115	217
24	479	11

degrees, respectively. The first trial model grid is oriented in the direction of flow in the Shallow Aquifer so that the model grid can be bounded by two flow lines (no-flow boundaries), one stream (constant head boundary), and one constant underflow boundary. Assume the impacts of sources and sinks within the model grid area will not appreciably affect these boundary conditions. A study of Shallow Aquifer thickness data for 24 wells (see Table 9) scattered throughout the first trial model grid area indicates that there are no other boundary conditions that need to be simulated.

Choose the "File" menu in the *DesignMod* Menu Bar and the "Exit" drop-down menu option. An Exit DesignMod message dialog box appears. Choose the "OK" button to exit *DesignMod*.

DM2 — Plan View

Exercise DM2 builds on the results of Exercise DM1 and helps you master the use of *DesignMod*'s File and Draw menu items. It also helps you master the Design Window Map tool. You will import a hydrogeologic well log database

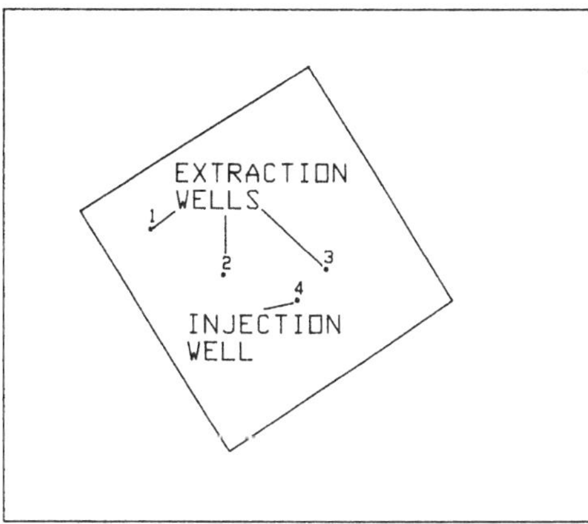

Figure 4.   Remediation scheme for Exercise DM2.

with 24 wells, draw a model plan view with site base map features and the first trial model grid borders defined in Exercise DM1, design the grid spacings, interpret water table elevations between wells, and overlay the plan view with the Shallow Aquifer water table contours to check the first trial model grid alignment.

The purpose of the modeling effort associated with Exercise DM2 is to define the possible impacts of a contaminant remediation scheme involving five extraction wells and one injection well developed in the Shallow Aquifer at locations within the first trial model grid borders, as shown in Figure 4.

Assume that a MOC PC (640K RAM) model (Konikow and Bredehoeft, 1984; Goode and Konikow, 1989) will be used in the modeling effort. MOC simulates two-dimensional contaminant migration in an upper layer (Shallow Aquifer) with vertical leakage to a lower layer (Deep Aquifer) through a Confining Bed represented by a vertical conductance between the two layers. Assume that the use of this model is justified because the Confining Bed and Deep Aquifer have not been contaminated and the Deep Aquifer is underlain by an aquiclude. Assume that Shallow Aquifer water levels and contaminant concentrations are depth-averaged for the modeling efforts. Model grid spacings are restricted because MOC PC limits the number of grid columns and rows to 20. Assume that MOC's steady-state mode is selected for the simulation.

In case you have MOC and would like to model these conditions, the horizontal hydraulic conductivity of the Shallow Aquifer ranges from 25 ft/day near well number 2 to 60 ft/day near well number 20. The horizontal hydraulic conductivity of the Deep Aquifer averages about 150 ft/day. The vertical hydraulic conductivity of the Confining Bed averages about 0.002 ft/day. The uniform annual recharge from precipitation rate is about 7 inches. The effective porosity is 0.25, the longitudinal dispersivity is 30 feet, and the transverse

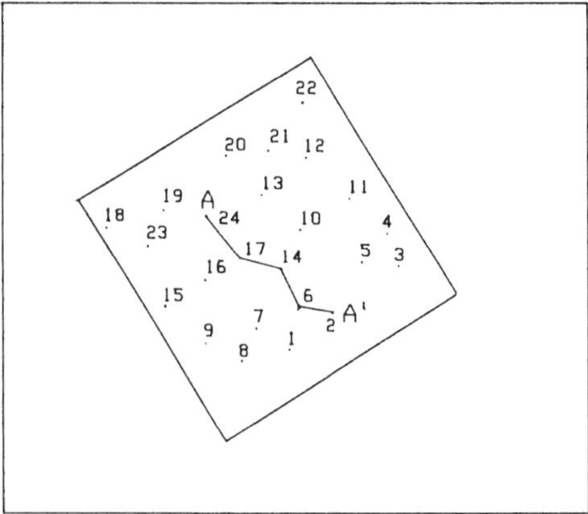

Figure 5.   Well locations for Exercise DM2.

dispersivity is 10 feet. The average flow rates in the Shallow Aquifer, Confining Bed, and Deep Aquifer are 200 ft/yr, 0.1 ft/yr, and 400 ft/yr, respectively.

The X- and Y-coordinates of the 24 wells in the database whose locations are shown in Figure 5 are given in Table 8.

Hydrogeologic logs for the 24 wells in the database are presented in Table 9. The number 999.99 signifies that no data are available.

There are three features on the site base map within the first trial model grid area: Elk Creek, Oak Street, and a building, as shown in Figures 1 and 6. Digitized X- and Y-coordinates for feature definition points are presented in Table 10.

The first trial model grid X- and Y-coordinate lower left corner offsets from the area of interest origin are 3300 and 1000 feet, respectively, as shown in Figure 7. The upper left and lower right first trial model grid corner X- and Y-coordinates are 1160, 4390 and 6660, 3100, respectively.

Start *DesignMod* and choose the "File" menu in the Manager Menu Bar. Select the "Load" drop-down menu option and the "Log" and "Hydrogeology" cascading menu options. A Specify Filename dialog box appears prompting for the name of the file in which the well log database is stored. The well log database file for exercise DM2 (based on data in Table 9) is stored in file DM2H.DAT on the disk. Type C:\DM\DM2H.DAT or a similar specification in the Filename text box, or choose the appropriate drive from the Drive List box and the appropriate directory from the Directory List box. Edit the Wildcard File List Extension text box to read *.DAT to limit the filenames displayed in the File List box. Choose the filename DM2H.DAT from the File List box.

Choosing the "OK" button stores the filename, closes the dialog box, and displays a Specify File Dimensions data entry dialog box prompting for information about the well and unit numbers for which well log data are stored in file DM2H.DAT. Type "24" in the Number Of Wells Or Models text box, "1"

Table 8. Well Coordinate Data for Exercise DM2

Well Number	X-Coordinate (ft)	Y-Coordinate (ft)
1	4230	2498
2	5044	2856
3	6150	3517
4	5710	4002
5	5303	3606
6	4378	2998
7	3720	2760
8	3560	2170
9	3015	2532
10	4508	4023
11	5230	4614
12	4680	5218
13	3770	4668
14	4023	3505
15	2460	3001
16	3020	3417
17	3685	3772
18	1700	4100
19	2501	4380
20	3200	5000
21	3927	5290
22	4505	5997
23	2222	3819
24	3002	4397

in the Well Or Model Number text box, and "3" in the Number Of Units Or Layers text box. Choose the "Enter Well Or Model" button to store these entries. Type "1" in the Unit or Layer Number text box. Choose the "Enter Unit Or Layer" button to store this entry. Choose the "Next Unit Or Layer" button to clear and reset the focus on the Unit Or Layer Number text box. Type "2" in the Unit Or Layer Number text box. Choose the "Enter Unit Or Layer" button to store this entry. Choose the "Next Unit Or Layer" button to clear and reset the focus on the Unit Or Layer Number text box. Repeat this process until all three unit numbers have been entered. Choose the "Next Well Or Model" button to clear text boxes and reset the focus on the Well Or Model Number text box. Enter data from Table 9 for well numbers 2 through 24 in the manner previously described. Be careful when you enter data for well number 17. Data are available for only one unit in that well, whereas data are available for three units in other wells. Choose the "OK" button to import the file, close the dialog box, and return to the Manager.

Table 9. Well Logs for Exercise DM2

Hydrogeologic Unit	Top Elevation (ft)	Base Elevation (ft)	Thickness (ft)	Water Level Elevation (ft)
Well Number 1				
Shallow Aquifer	562	543	19	562.18
Confining Bed	543	517	26	999.99
Deep Aquifer	517	492	25	552.33
Well Number 2				
Shallow Aquifer	562	545	17	562.23
Confining Bed	545	515	30	999.99
Deep Aquifer	515	492	23	552.44
Well Number 3				
Shallow Aquifer	562	545	17	562.30
Confining Bed	545	502	43	999.99
Deep Aquifer	502	492	10	553.82
Well Number 4				
Shallow Aquifer	561	541	20	561.32
Confining Bed	541	499	42	999.99
Deep Aquifer	499	494	5	555.07
Well Number 5				
Shallow Aquifer	562	542	20	561.55
Confining Bed	542	499	43	999.99
Deep Aquifer	499	493	6	554.08
Well Number 6				
Shallow Aquifer	562	541	21	561.71
Confining Bed	541	516	25	999.99
Deep Aquifer	516	494	22	552.06
Well Number 7				
Shallow Aquifer	562	540	22	561.62
Confining Bed	540	510	30	999.99
Deep Aquifer	510	493	17	552.82
Well Number 8				
Shallow Aquifer	562	543	22	562.03
Confining Bed	543	520	23	999.99
Deep Aquifer	520	492	28	552.47

Table 9 (continued). Well Logs for Exercise DM2

Hydrogeologic Unit	Top Elevation (ft)	Base Elevation (ft)	Thickness (ft)	Water Level Elevation (ft)
Well Number 9				
Shallow Aquifer	561	535	26	561.08
Confining Bed	535	520	15	999.99
Deep Aquifer	520	493	27	552.98
Well Number 10				
Shallow Aquifer	560	527	29	560.02
Confining Bed	527	497	30	999.99
Deep Aquifer	497	494	3	555.27
Well Number 11				
Shallow Aquifer	560	532	28	559.75
Confining Bed	532	498	34	999.99
Deep Aquifer	498	494	4	555.38
Well Number 12				
Shallow Aquifer	557	521	36	557.43
Confining Bed	521	497	24	999.99
Deep Aquifer	497	494	3	555.79
Well Number 13				
Shallow Aquifer	558	517	41	557.81
Confining Bed	517	497	20	999.99
Deep Aquifer	497	495	2	555.66
Well Number 14				
Shallow Aquifer	560	525	35	560.28
Confining Bed	525	499	26	999.99
Deep Aquifer	499	494	5	554.62
Well Number 15				
Shallow Aquifer	559	527	32	559.01
Confining Bed	527	508	19	999.99
Deep Aquifer	508	494	14	555.14
Well Number 16				
Shallow Aquifer	559	526	33	559.33
Confining Bed	525	502	24	999.99
Deep Aquifer	502	494	8	555.21

Table 9 (continued). Well Logs for Exercise DM2

Hydrogeologic Unit	Top Elevation (ft)	Base Elevation (ft)	Thickness (ft)	Water Level Elevation (ft)
Well Number 17				
Shallow Aquifer	560	521	39	559.53
Well Number 18				
Shallow Aquifer	555	521	34	555.21
Confining Bed	521	498	23	999.99
Deep Aquifer	498	495	3	555.88
Well Number 19				
Shallow Aquifer	556	523	33	556.08
Confining Bed	523	498	25	999.99
Deep Aquifer	498	495	3	555.82
Well Number 20				
Shallow Aquifer	556	517	39	555.73
Confining Bed	517	497	20	999.99
Deep Aquifer	497	495	2	556.25
Well Number 21				
Shallow Aquifer	556	518	38	556.19
Confining Bed	518	497	21	999.99
Deep Aquifer	497	495	2	555.93
Well Number 22				
Shallow Aquifer	556	519	37	555.66
Confining Bed	519	497	22	999.99
Deep Aquifer	497	495	2	556.95
Well Number 23				
Shallow Aquifer	557	524	33	557.01
Confining Bed	524	498	26	999.99
Deep Aquifer	498	495	3	555.68
Well Number 24				
Shallow Aquifer	557	525	32	557.12
Confining Bed	525	498	27	999.99
Deep Aquifer	498	495	3	555.38

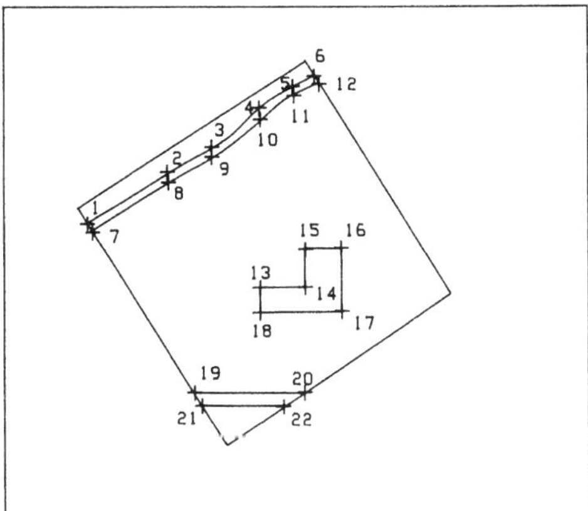

Figure 6.   Digitized site base map feature points for Exercise DM2.

Choose the "Draw" menu in the Manager Menu Bar and select the "Plan View" option in the drop-down menu. A Specify Plan View Dimensions data entry dialog box appears prompting for information about the site X- and Y-coordinates of the first trial model grid and displaying the block number 1. Type "1160" in the Upper Left Model Grid Corner Site X-Coordinate text box, "4390" in the Upper Left Model Grid Corner Site Y-Coordinate text box, "6660" in the Lower Right Model Grid Corner X-Coordinate text box, "3100" in the Lower Right Model Grid Corner text box, "FEET" in the X and Y Coordinate Unit text box, and "32" in the Rotation Angle Degree text box. Select the Counterclockwise Rotation Angle option. Type "3300" in the X-Axis Offset From Lower Left Model Grid Corner text box and "1000" in the Y-Axis Offset From Lower Left Model Grid Corner text box. Choose the "Enter" button to store these entries. Be aware that incorrect entries can result in unsuspected locations for Plan View Design Window site base map features and the model grid. Choose the "OK" button to close the dialog box and display the Plan View Design Window. Type "1" in the top center Block Number text box, and choose the adjacent "Enter" button to set the Plan View Design Window scale.

The next task is to draw site base map features on the Plan View Design Window. Start by drawing Elk Creek borders. Accept the Keyboard default drawing mode. Choose the "Width" tool to change the line width so that the Elk Creek borders will be thicker than the first trial model grid lines. A Select Line Width data entry dialog box appears with an optional Width List box. Select the "2 Pixel" option, and choose the "OK" button to close the dialog box and return to the Plan View Design Window. Choose the "Color" tool to change the line color from default black to blue so that Elk Creek borders will be blue. A Select Color dialog box appears with an optional Color List box. Select "Blue",

Table 10. Digitized Site Base Map Features for Exercise DM2

Point No.	X-Coordinate (ft)	Y-Coordinate (ft)
Elk Creek		
1	1350	4100
2	2170	4700
3	2770	5100
4	3700	5680
5	4200	5950
6	4700	6270
7	1390	4030
8	2170	4600
9	2770	5000
10	3700	5600
11	4200	5890
12	4730	6200
Building		
13	4000	3450
14	4700	3450
15	4700	4000
16	5200	4000
17	5200	3100
18	4000	3100
Oak Street		
19	2820	1800
20	4560	1800
21	2860	1700
22	4400	1700

and choose the "OK" button to close the dialog box and return to the Plan View Design Window.

Choose the "Curve" tool to draw the borders of Elk Creek on the Plan View Design Window. A Specify Curve Dimensions data entry dialog box appears prompting for information about curve point coordinates and displaying the point number 1. Draw the Elk Creek border delimited by points 1–6 first. Select the "Site Coordinate Type" option. Type "6" in the Number Of Points text box, "1350" in the Site X-Coordinate text box, and "4100" in the Y-Coordinate text box. Choose the "Enter" button to store these entries. Choose the "Next" button to clear text boxes and reset the focus so that data for point number 2 can be entered. Type "2170" in the Site X-Coordinate text box and "4700" in the Y-Coordinate text box. Choose the "Enter" button to

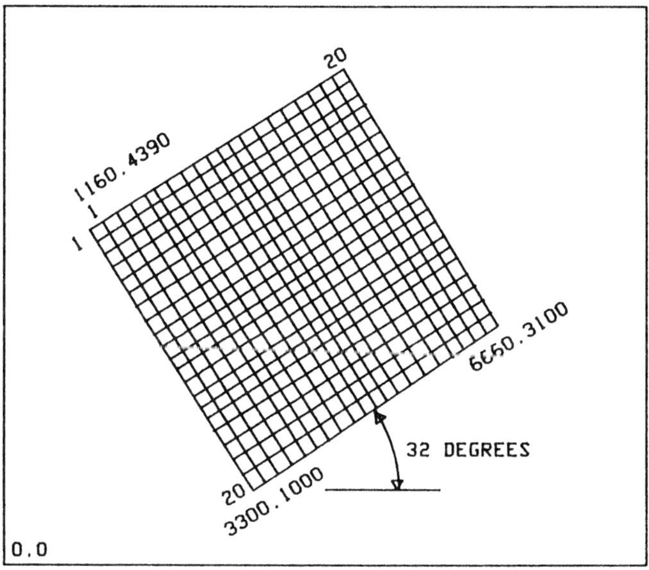

Figure 7.   First trial model grid coordinates for Exercise DM2.

store these entries. Continue this process until data for all six points have been entered. Choose the "Draw" button to draw the Elk Creek border delimited by points 1–6 in the picture box. Be patient; it takes a little while to interpret between control points and closely spaced pixel screen locations. Choose the "OK" button to close the dialog box and return to the Plan View Design Window. Set the line width and color as described earlier to 2 pixels and blue. Choose the "Curve" tool to draw the Elk Creek border delimited by points 7–12. The same process used in drawing the Elk Creek border delimited by points 1–6 is used to draw the Elk Creek border delimited by points 7–12. Choose the "OK" button to close the dialog box and return to the Plan View Design Window.

The next task is to draw the building borders on the Plan View Design Window. Choose the Width tool to change the line width so that the building outline will be thicker than first trial model grid lines. A Select Line Width data entry dialog box appears with an optional Width List box. Select the "2 Pixel" option, and choose the "OK" button to close the dialog box and return to the Design Window. Choose the "Color" tool to change the line color from default black to green so that the building outline will be green. A Select Color dialog box appears with an optional Color List box. Select "Green", and choose the "OK" button to close the dialog box and return to the Plan View Design window.

Choose the "Poly" tool to draw the building outline. A Specify Polygon Dimensions data entry dialog box appears prompting for information about the coordinates of the building outline vertices and displaying vertice number 1.

Select the "Site Coordinate Type" option, and type "6" in the Number Of Vertices text box. Type "4000" in the Site X-Coordinate text box and 3450 in the Site Y-Coordinate text box. Choose the "Enter" button to store these entries. Choose the "Next" button to enter data for vertex number 2. Type "4700" in the X-Coordinate text box and "3450" in the Y-Coordinate text box. Choose the "Enter" button to store these entries. Choose the "Next" button to enter data for additional vertices. Continue this process until data for all six vertices are entered. Choose the "Draw" button to draw the building borders. Choose the "OK" button to close the dialog box and return to the Plan View Design Window.

The next task is to draw the Oak Street borders on the Plan View Design Window. Choose the "Width" tool to change the line width so that the Oak Street border from point 19 to point 20 will be thicker than first trial model grid lines. A Select Line Width data entry dialog box appears with an optional Width List box. Select the "2 Pixel" option, and choose the "OK" button to close the dialog box and return to the Plan View Design Window. Choose the "Color" tool to change the line color from default Black to Cyan so that Oak Street borders will be Cyan. A Select Color dialog box appears with an optional Color List box. Select "Cyan", and choose the "OK" button to close the dialog box and return to the Plan View Design window.

Choose the Line tool to draw the Oak Street border from point 19 to point 20. A Specify Line Dimensions data entry dialog box appears prompting for information about the line definition points. Select the "Site Coordinate Type" option. Type "2820" in the Site Start X-Coordinate text box, "1800" in the Site Start Y-Coordinate text box, "4560" in the Site End X-Coordinate text box, and "1800" in the Site End Y-Coordinate text box. Choose the "Draw" button to store these entries and draw the Oak Street border from point 19 to point 20. Choose the "OK" button to close the dialog box and return to the Plan View Design Window.

Choose the "Width" tool to change the line width so that the other Oak Street border will be thicker than first trial model grid lines. A Select Line Width data entry dialog box appears with an optional Width List box. Select the "2 Pixel" option, and choose the "OK" button to close the dialog box and return to the Design Window. Choose the "Color" tool to change the line color from default Black to Cyan so that Oak Street borders will be Cyan. A Select Color dialog box appears with an optional Color List box. Select "Cyan", and choose the "OK" button to close the dialog box and return to the Plan View Design window.

Choose the "Line" tool to draw the Oak Street border from point 21 to point 22. A Specify Line Dimensions data entry dialog box appears prompting for information about the line definition points. Select the "Site Coordinate Type" option. Type "2860" in the Site Start X-Coordinate text box, "1700" in the Site Start Y-Coordinate text box, "4400" in the Site End X-Coordinate text box, and "1700" in the Site End Y-Coordinate text box. Choose the "Draw" button to store these entries and draw the Oak Street border from point 21 to point 22. Choose the "OK" button to close the dialog box and return to the Plan View Design window.

The next task is to label the site base map features. Choose the "Font" tool to label Elk Creek, the building, and Oak Street. A Select Font Attributes dialog box appears offering font name, size, and color options. Select "Helv" from the Name List box, "9.75" from the Size List box, and "Cyan" from the Color List box. Choose the "Enter" button to enter these choices. Sample text appears in the text box. Choose the "OK" button to close the dialog box and return to the Plan View Design Window.

Use the mouse cursor to locate appropriate points for the upper left corners of the labels. Make a note of the X- and Y-coordinates of the points appearing at the top left corner of the Plan View Design Window (1635, 430 for Elk Creek; 3036, 2027 for the building; and 1020, 3427 for Oak Street). Choose the "Text" tool to label Elk Creek. A Specify Text dimensions data entry dialog box appears prompting for the upper left corner coordinates of the text string label. Select the "Screen Coordinate Type" option. Type "1635" in the Screen X-Coordinate text box, "430" in the Screen Y-Coordinate text box, and "ELK CREEK" in the Text box. Choose the "Enter" button to store these entries and label Elk Creek. Choose the "Next" button to label the building. Type "3036" in the Screen X-Coordinate text box, "2027" in the Screen Y-Coordinate text box, and "BUILDING" in the Text box. Choose the "Enter" button to store these entries and label the building. Choose the "Next" button to label Oak Street. Type "1020" in the Screen X-Coordinate text box, "3427" in the Screen Y-Coordinate text box, and "OAK STREET" in the Text box. Choose the "Enter" button to store these entries and label Oak Street. Choose the "OK" button to close the dialog box and return to the Plan View Design Window.

Choose the "Grid" tool to select the first trial model grid spacings and draw first trial model grid lines on the plan view. A Specify Grid Spacing Dimensions data entry dialog box appears prompting for information about grid spacings and column and row numbers and displaying the block number 1. Assume that, in light of well locations and other considerations, a 20 by 20 grid with 200-ft spacing seems reasonable for the first trial model grid spacings. Type "1" in the Block Column Data First Number text box, "20" in the Block Column Data Last Number text box, and "200" in the Block Column Grid Spacing text box. Choose the "Enter Col" button to store these entries, and choose the "Row" button to reset the focus on block row. Type "1" in the Block Row Data First Number text box, "20" in the Block Row Data Last Number text box, and "200" in the Block Row Grid Spacing text box. Choose the "Enter Row" button to store these entries. Type "1" in the Grid Data Left Column Number text box, "20" in the Grid Data Right Column Number text box, "1" in the Grid Data Upper Row Number text box, and "20" in the Grid Data Lower Row Number text box. Choose the "Enter Grid" button to store these entries and draw the first trial model grid on the Plan View Design Window. Choose the "OK" button to close the dialog box and return to the Plan View Design Window.

Choose the "Map" tool to overlay the Shallow Aquifer water table contours on the plan view to check the first trial model grid alignment. A Map menu appears. Choose the "Data" menu in the Map Menu Bar, the "Surface" drop-down menu item, and the "Well" cascading menu option to identify the wells and units whose X- and Y-coordinates and water level data will be used to create a gridded water table surface array with the Interpret menu. A Specify Surface Well Dimensions data entry dialog box appears prompting for well and unit information. Select the "Hydrogeologic Log Type" and the "Water Level Surface Type" options. Type "24" in the Number Of Wells To Be Used In Interpretation text box, "1" in the Well Number text box, and "1" in the Unit Number text box. Choose the "Enter" button to store these entries and type choices. Choose the "Next" button to clear text boxes so that data for well number 2 can be entered. Type "2" in the Well Number text box and "1" in the Unit Number text box. Choose the "Enter" button to store these entries. Repeat this process until data for all 24 wells have been entered. Choose the "OK" button to close the dialog box and return to the Map menu.

Choose the "File" menu in the Map Menu Bar, the "Save" drop-down menu, and the Scattered Triplet cascading menu option to save scattered water table data for unit number 1 in a file. A Specify Well Or Point Dimensions dialog box is displayed. Select the "Wells Data Type" option, type "24" in the Number Of Wells text box, and type a valid filename such as C:\DM\DM2S.DAT in the Filename text box. Choose the "Enter" button to store these entries. Choose the "OK" button to save water table data for unit 1 in a scattered triplet array file with lower left origin and site base map coordinates.

Choose the "Interpret" menu in the Map Menu Bar and the "Surface" drop-down menu item to interpret water table elevations between the 24 wells and create a first trial model grid array of Shallow Aquifer water table elevations. A Specify Interpretation Dimensions dialog box appears offering interpretation technique options. Select the "Triangulation Interpretation Technique" option. Choosing the "OK" button creates the first trial model grid array, closes the dialog box, and returns to the Map menu.

Choose the "File" menu in the Map Menu Bar, the "Save" drop-down menu item, and the "Gridded Triplet" cascading menu option to save the Shallow Aquifer water table elevation array in a file. A Specify Filename dialog box is displayed. Type a valid filename such as C:\DM\DM2G.DAT in the Save As text box. Choose the "Enter" button to store this entry. Choose the "OK" button to save the Shallow Aquifer water table data in a gridded triplet array file with upper left origin and model coordinates.

Choose the "Overlay" menu in the Map Menu Bar, the "Contour" drop-down menu item, and the "Screen" cascading menu option to overlay the plan view with the Shallow Aquifer water table contours as shown in Figure 8.

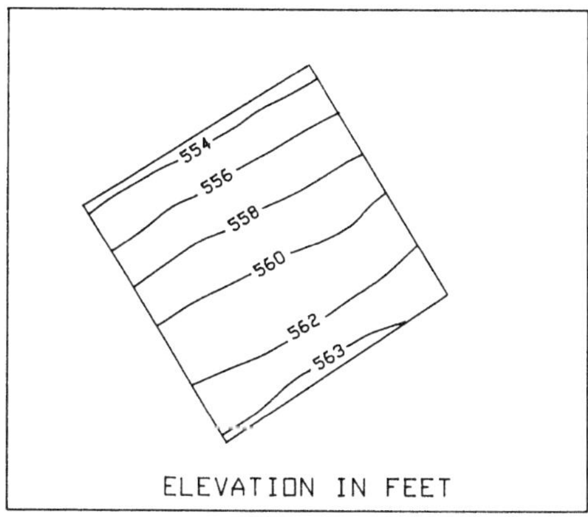

Figure 8. Water table contours for Exercise DM2.

A Contour Map Dimensions dialog box appears prompting for information about contours and displaying contour number 1. Type "5" in the Number Of Contours text box and "557" in the Value text box. Choose the "Enter" button to store these entries. Choose the "Next" button to clear the Value text box and enter data for contour number 2. Type "558" in the Value text box and choose the "Enter" button to store this entry. Repeat this process for contour numbers 3, 4, and 5, whose values are 559, 560, and 561, respectively. Choose the Plot button to draw the Shallow Aquifer water table contours on the plan view. Be patient; this process may take awhile. Choose the "OK" button to close the dialog box and return to the Map menu. Choose the "File" menu in the Map Menu Bar and the "Exit" drop-down menu option to close the Map menu and return to the Plan View Design Window.

Choose the "Legend" tool to determine the values associated with the colored contours. A Legend Option dialog box appears offering five Type options — Zone, Cell, Info, Contour, and Zonal Diagram — and two Switch options, On and Off. Select the "Contour" Type option, the "On Switch" option, and the "OK" button. Position the mouse cursor on top of any contour and depress the left mouse button. The contour value is displayed at the lower left corner of the Plan View Design Window. Choose the "Legend" tool, "Contour" Type option, "Off Switch" option, and "OK" button to disable the Legend tool.

The first trial model grid alignment seems to be correct because the grid borders are at right angles to the Shallow Aquifer water table contours, as they should be. Choose the "Undo" tool to erase the contours and unclutter the Plan View Design Window.

MOC automatically specifies the outer rows and columns of the model flow finite-difference grid as no-flow boundaries. Assume that it has been deemed appropriate to represent the flow lines at right angles to the Shallow Aquifer water table contours along columns 1 and 20 as no-flow boundaries because the remedial scheme is not expected to affect these flow lines. Identify these no-flow boundaries on the plan view by choosing the "Cell" tool. A Specify Cell Block Dimensions data entry dialog box appears prompting for information about cell block color, description, and dimensions and displaying the type number 1 and the block number 1. Select "Black" in the Color List box. Type "NO-FLOW BOUNDARY" in the Description text box. Type "1" in the Upper Left Corner Column and Row Number text boxes, "20" in the Lower Left Corner Column Number text box, and "1" in the Lower Left Corner Row Number text box. Choose the "Enter" button to store these entries. Choose the "Draw" button to color the cells within the specified block number 1 dimensions. Choose the "Next" Block button to clear text boxes and increment the block number so that dimensions for block number 2 can be entered.

Type "1" in the Upper Left Corner Column Number text box, "20" in the Upper Left Corner Row Number text box, "20" in the Lower Left Corner Column Number text box, and "20" in the Lower Left Corner Row Number text box. Choose the "Enter" button to store these entries. Choose the "Draw" button to color the cells within the specified block number 2 dimensions black. The block number 2 description will be No-Flow Boundary. Choose the "Next Block" button to increment the block number and clear text boxes so that dimensions for block number 3 can be entered.

Type "1" in the Upper Left Corner Column Number text box, "2" in the Upper Left Corner Row Number text box, "1" in the Lower Left Corner Column Number text box, and "19" in the Lower Left Corner Row Number text box. Choose the "Enter" button to store these entries. Choose the "Draw" button to color the cells within the specified block number 3 dimensions black. The block number 3 description will be No-Flow Boundary. Choose the "Next Block" button to increment the block number and clear text boxes so that dimensions for block number 4 can be entered.

Type "20" in the Upper Left Corner Column Number text box, "2" in the Upper Left Corner Row Number text box, "20" in the Lower Left Corner Column Number text box, and "19" in the Lower Left Corner Row Number text box. Choose the "Enter" button to store these entries. Choose the "Draw" button to color the cells within the specified block number 4 dimensions black. The block number 4 description will be No-Flow Boundary. Choose "Next Type" to increment the type number, set the block number to 1, and clear text boxes so that constant head cells can be identified.

Assume that heads along rows 2 and 19 are not expected to be affected by the remedial scheme. Elk Creek can be represented by a constant head (554 feet) boundary along row 2, and the head along row 19 can be set at a constant value of 563 feet. Select "Blue" in the Color List box. Type "CONSTANT HEAD" in the Description text box. Type "2" in the Upper Left Corner Column Number text box, "2" in the Upper Left Corner Row Number text box, "19" in the Lower Left Corner Column Number text box, and "2" in the Lower Left Corner Row Number text box. Choose the "Enter" button to store these entries. Choose the "Draw" button to color the cells within the specified block number 1 dimensions Blue. Choose the "Next Block" button to increment the block number and clear text boxes so that dimensions for block number 2 can be entered.

Type "2" in the Upper Left Corner Column Number text box, "19" in the Upper Left Corner Row Number text box, "19" in the Lower Left Corner Column Number text box, and "19" in the Lower Left Corner Row Number text box. Choose the "Enter" button to store these entries. Choose the "Draw" button to color the cells within the specified block number 2 dimensions blue. The block number 2 description will be "CONSTANT HEAD BOUNDARY". Choose the "OK" button to close the dialog box and return to the Plan View Design Window.

The border no-flow boundaries and constant head boundaries along rows 2 and 19 within the no-flow boundaries are displayed. Descriptions for the boundaries can be viewed by choosing the "Legend" tool, selecting the "Cell" Type and "On Switch" option in the Legend Options dialog box which appears, choosing the "OK" button to return to the Plan View Design Window, moving the mouse cursor over a color cell, and pressing the left mouse button. The boundary description is displayed below the picture box in the Plan View Design Window.

The remedial scheme consists of three extraction wells and one injection well at locations shown in Figure 4. Well coordinates are listed in Table 11.

Choose the "Info" tool to check the first trial model grid alignment and spacing with respect to the remedial scheme. A Specify Information Point Dimensions data entry dialog box appears prompting for the extraction and injection well point X- and Y-coordinates. The type number 1 and the point number 1 are displayed. Select the "Site Coordinate" Type option and "Green" in the Color Type List box to represent the extraction wells. Type "2300" in the Site X-Coordinate text box, "4300" in the Site Y-Coordinate text box, and "EXTRACTION WELL" in the Description text box. Choose the "Enter" button to store these entries and the "Draw" button to draw a small green-colored circle at the location of extraction well 1. Choose the "Next Point" button to increment the point number, clear text boxes, and reset the focus so that data for extraction well 2 can be entered. Type "3300" in the Site X-Coordinate text box and "3500" in the Site Y-Coordinate text box. Choose the "Enter" button to store these entries and the "Draw" button to draw a small green-colored circle at the location of extraction well 2. The point number 2 description will be "EXTRACTION WELL". Choose the "Next Point" button to increment the point number, clear text boxes, and reset the focus so that data for extraction well 3 can be entered. Type "4800" in the Site X-Coordinate text box and "3700" in the Site Y-Coordinate text box. Choose the "Enter" button to store these entries and the "Draw" button to draw a small green-colored

Table 11. Digitized Site Extraction and Injection Well Coordinates for Exercise DM2

Well Number	X-Coordinate (ft)	Y-Coordinate (ft)
Extraction		
1	2300	4300
2	3300	3500
3	4800	3700
Injection		
4	4150	3300

circle at the location of extraction well 3. The point number 3 description will be "EXTRACTION WELL".

Choose the "Next Type" button to increment the type number, set the point number to 1, and enter data for the injection well. Select the "Site Coordinate" option and "Magenta" in the Color List box. Type "4150" in the Site X-Coordinate text box, "3300" in the Y-Coordinate text box, and "INJECTION WELL" in the Description text box. Choose the "Enter" button to store these entries and the "Draw" button to draw a small magenta-colored circle at the location of the injection well. Choose the "OK" button to close the dialog box and return to the Plan View Design Window.

Choose the "Legend" tool to view the information point legend. A Legend Option dialog box appears offering five Type options — Zone, Cell, Info, Contour, and Zonal diagram — and two Switch Options, On and Off. Select the "Info Type" and "On Switch" options. Choose the "OK" button to close the dialog box and return to the Plan View Design window. Position the mouse cursor on a small green circle and depress the left mouse button. The information type number 1 description Extraction Well is displayed below the picture box.

The model grid alignment and spacing seem to be reasonable with respect to the extraction and injection well locations. The first trial model grid alignment and spacing are judged to be reasonable based on the Shallow Aquifer water table and remedial scheme analyses described above.

Choose the "OK" tool to close the Design Window. Choose the "File" menu in the *DesignMod* Menu Bar, and the "Exit" drop-down menu option. An Exit DesignMod message dialog box appears. Choose the "OK" button to exit *DesignMod*.

DM3 — Zonal Diagram

This exercise helps you master the Edit, View, and File menus and associated dialog boxes as you create and save a gridded triplet array with

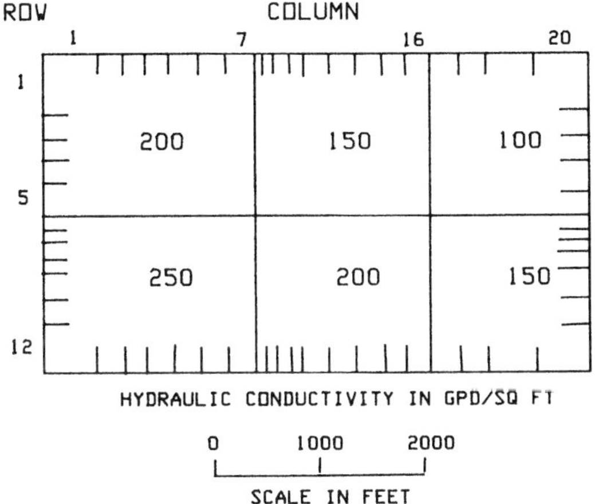

Figure 9.   Z value zone blocks for Exercise DM3.

variable grid spacing. The array grid has 20 columns and 12 rows and is subdivided into 6 Z value zone blocks as shown in Figure 9. The Z values represent aquifer hydraulic conductivities. The gridded triplet array is stored in file DM3G.DAT on the disk. Block Z values are listed in Table 12.

Column and row grid spacings are listed in Table 13. The X- and Y-coordinates of the array upper left grid node are each 500.

Start *DesignMod* and choose the "Edit" menu in the Manager Menu Bar. Select the "Z" or "XYZ Array" drop-down menu option and the "New" cascading menu option. A Specify Edit Block Dimensions dialog box appears prompting for information about zone block dimensions, the array type, and the zone block Z value. The number 1 appears in the Block Number text box. Type "1" in the Upper Left Corner Column and Row Number text boxes. Type "7" in the Lower Right Corner Column Number text box, and "5" in the Lower Right Corner Row Number text box. Select the "Gridded Triplet Array" and the "Edit Grid Spacing" options. Type "200" in the Value Or Operator Factor text box. Choose the "Enter" button to store entries for block number 1.

Table 12. Z Value Data for Exercise DM3

Block Number	Columns	Rows	Z Value (gpd/sq ft)
1	1–7	1–5	200
2	8–16	1–5	150
3	17–20	1–5	100
4	1–7	6–12	250
5	8–16	6–12	200
6	17–20	6–12	150

Table 13. Grid Spacing Data for Exercise DM3

First Column Number	Last Column Number	Grid Spacing (ft)
1	1	1000
2	7	500
8	11	250
12	18	500
19	20	1000

First Row Number	Last Row Number	Grid Spacing (ft)
1	1	1000
2	5	500
6	9	250
10	11	500
12	12	1000

Choose the "Next" button to clear text boxes, reset the focus, and increment the block number. Since you have already selected the Array Type and Grid Spacing options, those options need not be changed. Type "8" in the Upper Left Corner Column, and "1" in the Upper Left Corner Row Number text boxes. Type "16" in the Lower Right Corner Column Number text box, and "5" in the Lower Right Corner Row Number text box. Type 150 in the Value Or Operator Factor text box. Choose the "Enter" button to store these entries and the "Next" button to clear text boxes and reset the focus so that data for additional blocks can be entered. Repeat the process described above until data for all six blocks have been entered. After you have successfully entered values for all blocks, choose the "OK" button to close the dialog box.

A Specify Grid Spacing Dimensions data entry dialog box appears prompting for information about grid spacings and displaying the block number 1. Type "1" in the Block Column Data First Number and Last Number text boxes. Type "1000" in the Block Column Data Grid Spacing text box. Choose the "Enter Column" button to store these entries. Choose the "Next Column" button to clear text boxes and reset the focus so that spacings for other columns can be entered. Type "2" in the Block Column Data First Number text box, and "7" in the Block Column Data Last Number text box. Type "500" in the Block Column Data Grid Spacing text box. Choose the "Enter" column button to store these entries and the "Next Column" button to clear text boxes and reset the focus. Repeat this process until data for all five column grid spacing blocks have been entered. Choose the "Row" button to reset the focus on Row text boxes.

Type "1" in the Block Row Data First Number and Last Number text boxes. Type "1000" in the Block Row Data Grid Spacing text box. Choose the "Enter Row" button to store these entries. Choose the "Next Row" button to clear text boxes and reset the focus so that grid spacings for other rows can be entered. Type "2" in the Block Row Data First Number text box, and "5" in the Block Row Data Last Number text box. Type "500" in the Block Row Data Grid Spacing text box. Choose the "Enter" column button to store these entries and the "Next Row" button to clear text boxes and reset the focus. Repeat this process until data for all five row grid spacing blocks have been entered.

Type "500" in the Grid Upper Left Node X-Coordinate and Upper Left Node Y-Coordinate text boxes. Type "1" in the Grid Left Column Number text box, "20" in the Grid Right Column Number text box, "1" in the Grid Upper Row Number text box, and "12" in the Grid Lower Row Number text box. Notice that these dimensions are for the entire array grid consisting of one or more zone blocks, whereas the previous dimensions were those for individual column and row blocks. Choose the "Enter Grid" button to store these entries and create the gridded triplet array with an upper left origin and a variable grid spacing. Choose the "OK" button to close the dialog box and return to the Manager.

You can quickly verify any portion of the array or all of the array in one or more viewport blocks as a screen and/or printer zonal diagram. Choose the "View" menu in the Manager Menu Bar, the "Map" drop-down menu option, and the "Zonal Diagram" and "Screen" cascading menu options to view the entire array. A Specify View Block Dimensions dialog box appears prompting for information about the viewport block dimensions, upper left block corner grid spacings, and XY-coordinate and Z value units. Notice that the upper left corner column and row grid spacings are requested, not the upper left corner X- and Y-coordinates. The number 1 appears in the Block Number text box. Type "1" in the Grid Upper Left Corner Column and Row Number text boxes. Type "20" in the Grid Lower Right Corner Column Number text box and "12" in the Grid Lower Right Corner Row Number text box. Type "1000" in the Upper Left Grid Spacing Column and Row text boxes. Type "FEET" in the XY-Coordinate Unit text box and "GPD/SQ FT" in the Z Value Unit text box. Choose the "Enter" button to store these entries. Choose the "OK" button to close the dialog box.

A Zonal Diagram Display picture and dialog box appears prompting for information about the block to be viewed and zonal diagram Z value ranges and displaying the X- and Y-coordinate and Z value units and the range number 1. Type "1" in the Block Number text box. Type "250" in the $\geq$ Z Value Range text box, and "251" in the $<$ Z Value Range text box. Choose the "Enter" button to store these entries. The following information appears in the picture box: Block Upper Left Column = 1, Block Upper Left Row = 1, Block Lower Right Column = 20, Block Lower Right Row = 12, Minimum Z Value = 100, Minimum Z Value Column = 17, Minimum Z Value Row = 2, Maximum Z Value = 250, Maximum Z Value Column = 2, and Maximum Z Value Row = 6. Choose the "Next" button to enter another range. Type "200" in the $\geq$ Range Value text box, and "201" in the $<$ Range Value text box. Choose the "Enter" button to store these entries and the "Next" button to clear text boxes and reset the focus. Repeat this process for range number 3 with a $150 \geq$ range value and $151 <$ range value, and for range number 4 with a $100 \geq$ range value and a $101 <$ range value. Choose the "Plot" button to erase information in the picture box and display the zonal diagram in the picture box. Choose the "Grid" button to display the model cell border line grid on top of the zonal diagram. Compare data in Table 12 with the zonal diagram displayed in the picture box to verify data entry.

Choose the "Legend" button to view the Z value ranges associated with any zonal diagram color. When you place the mouse cursor within any color block and depress the left mouse button, the Z value range appears above the upper left corner of the picture box.

You can view the column and row coordinates of cell nodes and/or the X- and Y-coordinates of any point on the zonal diagram by choosing the appropriate Axes option, moving the mouse pointer to the desired node or point and depressing the left mouse button. As soon as the mouse button is depressed, the coordinates of the node or point appear in the Coordinate text boxes to the right of the picture box. Choose the "OK" button to close the dialog box and return to the Manager.

You can obtain a hard copy of the zonal diagram by choosing "View" in the Manager Menu Bar, selecting the "Map" drop-down menu option, and selecting the "Zonal Diagram" and "Printer" cascading options. A Specify View Block Dimensions dialog box appears prompting for information about the dimensions of the array block to be printed and the upper left grid spacings and displaying 1 in the Block Number text box. Type "1" in the Upper Left Corner Column and Row text boxes. Type "20" in the Lower Right Corner Column text box, and "12" in the Lower Right Corner Row text box. Type "1000" in the Upper Left Grid Spacing Column and Row text boxes. Choose the "Enter" button to store these entries. Choose the "OK" button to close the dialog box.

A Specify Printer Zonal Diagram Dimensions data entry dialog box appears prompting for information about the Z value ranges and the Z value unit. Type "1" in the Block Number text box. Type "250" in the $\geq$ Range Value and "251" in the $<$ Range Value text boxes. Type "GPD/FT" in the Z Value Unit text box. Choose the "Unit" button to store the unit entry. Choose the "Range" button to store the range values. Choose the "Next Range" button to enter additional range values. Type "200" in the $\geq$ Range Value text box, and "201" in the $<$ Range Value text box. Choose the "Range" button to store these entries. Choose the "Next Range" button to clear text boxes and reset the focus. Repeat this process until all four zones are defined. Choose the "Print" button to print the block number 1 zonal diagram. A letter is printed at each model cell node depending on the relation between the node Z value and the zone range values assigned to the letter. Model grid dimensions, Z value unit, and Z value range legend are printed below the zonal diagram. Invalid nodal values are easily identified by comparing data in Table 12 and the zonal diagram. Choose the "OK" button to close the dialog box and return to the Manager.

You can view the zonal diagram array XYZ values in a screen spreadsheet box by choosing the "View" menu in the Manager Menu Bar; selecting the "Tabular" drop-down menu option; and selecting the "Triplet", "Gridded", and "Screen" cascading menu options. A Specify Tabular Display Dimensions dialog box appears prompting for information about the block number, display type, viewport block dimensions, upper left node data, and X- and Y-coordinate and Z value units. Type "1" in the Block Number text box. Select the "XYZ Values Display" Type option. Type "1" in the Upper Left Corner Column and Row Number text boxes. Type "20" in the Lower Right Corner Column Number text box and 12 in the Lower Right Corner Row Number text box. Type "FEET" in the X- and Y-Coordinate Unit text boxes and "GPD/SQ FT" in the Z Value Unit text box. Choose the "Enter" button to store these entries.

Choose the "OK" button to close the dialog box. A Tabular Display spreadsheet and dialog box appears requesting information about the block number and the desired format of the values to be displayed. Type "1" in the Block Number text box and select the "Decimal" Format option. Choose the "Facts" button. The following information appears in an information message box: Upper Left Column = 1, Upper Left Row = 1, Lower Right Column = 20, Lower Right Row = 12, Minimum X-Value = 500, Minimum X-Column = 1, Maximum X-Value = 10000, Maximum X Column = 20, Minimum Y-Value = 500, Minimum Y Row = 1, Maximum Y-Value = 5500, Maximum Y Row = 12, Minimum Z Value = 100, Minimum Z Column = 17, Minimum Z Row = 1, Maximum Z Value = 250, Maximum Z Column = 1, and Maximum Z Row = 6. Choose the "OK" button to close the information message box.

Choose the "View" button, and the XYZ values for block number 1 appear in the spreadsheet box. Table headings are Row, Column, X-Coordinate, Y-Coordinate, and Z Value. X- and Y-coordinate and Z value units appear below the table headings. Use the spreadsheet box scroll bars or arrow keys to view values throughout the block. Choose the "OK" button to close the spreadsheet dialog box and return to the Manager.

You can obtain a hard copy of a block of the zonal diagram array XYZ values as a table by choosing the "View" menu in the Manager Menu Bar; selecting the "Tabular" drop-down menu option; and selecting the "Triplet", "Gridded", and "Printer" cascading options. A Specify Array Type dialog box appears. Select the "XYZ" option. Choose the "OK" button to close the dialog box. A Specify Printer Block Dimensions data entry dialog box appears prompting for information about the block dimensions, the X- and Y-coordinate and Z value units, and the numerical format to be used in printing values. Type "1" in the Upper Left Corner Column and Row Number text boxes. Type "20" in the Lower Right Corner Column Number text box and "12" in the Lower Right Corner Row Number text box. Type "FEET" in the X- and Y-Coordinate Unit text boxes and "GPD/SQ FT" in the Z Value Unit text box. Select the "Decimal" Format option. Choose the "Print" button to obtain a hard copy of the block of array X, Y, and Z values. Table headings are Row, Column, X-Coordinate, Y-Coordinate, and Z Value. The X- and Y-coordinate and Z value units appear below the headings. You can print values for another block by choosing the "Next" button. Choose the "OK" button to close the dialog box and return to the Manager.

Any portion of the newly created zonal diagram array or the entire newly created zonal diagram array can be saved as a grid or triplet file by specifying appropriate file array dimensions. To save the entire newly created zonal diagram array as a gridded triplet array with variable grid spacings and a lower left origin, choose the "File" menu in the Manager Menu Bar. Select the "Save" drop-down menu option and the "Gridded Triplet" and "Lower Origin" cascading menu options. A Specify File Dimensions data entry dialog box appears prompting for information about the file array dimensions and the name of the file to be saved. Type "1" in the Array Upper Left Corner Column and Row Number text boxes. Type "20" in the Array Lower Right Corner Column Number text box, and "12" in the Array Lower Right Corner Row Number text box. Type a valid name such as C:\DM\DM3G.DAT for the file in which the array is to be saved in the Save As text box. Choose the "Enter" button to create the file. Choose the "OK" button to close the dialog box and return to the Manager.

You can retrieve the saved array file and browse its contents by choosing the "File" menu and selecting the "Browse" drop-down menu option. A Filename dialog box appears prompting for the name of the file to be saved. Type a valid filename specification such as C:\DM\DM3G.DAT in the Filename text box and choose the "OK" button. You can also choose the appropriate drive and directory from the list boxes. The default Wildcard File List Extension *.* which displays all filenames in the selected directory can be edited. If you wish to display a group of filenames with a common extension such as *.DAT and not all filenames in the selected directory, type the desired wildcard extension in the Wildcard File List Extension text box. Choose the "Extension" button to activate the filename search and display. Select the filename from the File List box and choose the "OK" button.

A Browse File text and dialog box appears with the filename displayed at the top of the screen. The contents of the entire file can be viewed by using the text box scroll bars or arrow keys. Choose the "Value" button to obtain information about the number of values in the file. A File Fact dialog box appears announcing that the number of values in the file is 720. Choose the "OK" button to close the File Fact dialog box and return to the Browse File text and dialog box. Choose the "Size" button to obtain information about the size of the file. A File Fact dialog box appears announcing that the size of the file is 3580 bytes. Choose "OK" to close the File Fact dialog box and return to the Browse File text and dialog box. Choose the "OK" button to return to the Manager.

Choose the "File" menu in the Manager Menu Bar and select the "Exit" drop-down menu option. A message appears in an Exit DesignMod Message box. Choose the "OK" button to close *DesignMod* and return to the *Windows* Program Manager.

DM4 — Contour Map

This exercise helps you master Edit, File, View, and Interpret menus and associated dialog boxes as you create, view, and save a scattered triplet array and create, view, and save two gridded model cell node triplet arrays with variable grid spacings. The scattered triplet array is stored in file DM4S.DAT on the disk. One gridded model cell node triplet array (file DM4GT.DAT on the disk) is created with the scattered triplet array and the triangulation interpolation technique, and the other gridded model cell node triplet array (file DM4GK.DAT on the disk) is created with the scattered triplet array and the kriging interpolation technique. The scattered triplet array with an upper left origin has 15 control points as shown in Figure 10. The control point reference frame upper left row Y-coordinate is 1000 feet, and the control point reference frame lower row Y-coordinate is 5000 feet. The model cell node grid (interpolation grid) overlying the scattered control points has 10 columns and 10 rows within the control point reference frame. The model cell node grid upper left corner X- and Y-coordinates are each 1000 feet. The Z values represent aquifer water levels. X, Y, and Z data for the scattered triplet array are presented in Table 14. The model cell node grid spacings are presented in Table 15.

Start *DesignMod* and choose the "Edit" menu in the Manager Menu Bar. Select the "Scattered Triplet" drop-down menu option. A Specify Scattered Point Dimensions data entry dialog box appears, prompting for information about the scattered triplet array and displaying 1 in the Point Number text box. Type "1850" in the X-Coordinate text box, "1290" in the Y-Coordinate text box, and "801" in the Z Value text box. Choose the "Enter" button to store these entries and the "Next" button to clear text boxes and reset the focus so that values can be entered for point number 2. Type "2475" in the X-Coordinate text box, "1510" in the Y-Coordinate text box, and "797" in the Z Value text box. Choose the "Enter" button to store these entries and the "Next" button to clear text boxes and reset the focus so that values for point number 3 can be entered. Repeat this process until values for all 15 control points have been entered. Choose the "OK" button to close the dialog box and return to the Manager.

Choose the "View" menu in the Manager Menu Bar; select the "Tabular" drop-down menu option; and select the "Triplet", "Scattered", and "Screen" cascading options to view the entire scattered triplet array. A Specify Tabular Display data entry dialog box appears, prompting for information about the viewport block dimensions. Type "1" in the Block Number and the First Point Number text boxes. Type "15" in the Last Point Number text box; and "FEET" in the X- and Y-Coordinate Unit and Z Value Unit text boxes. Choose the "Enter" button to store these entries. Choose the "OK" button to close the

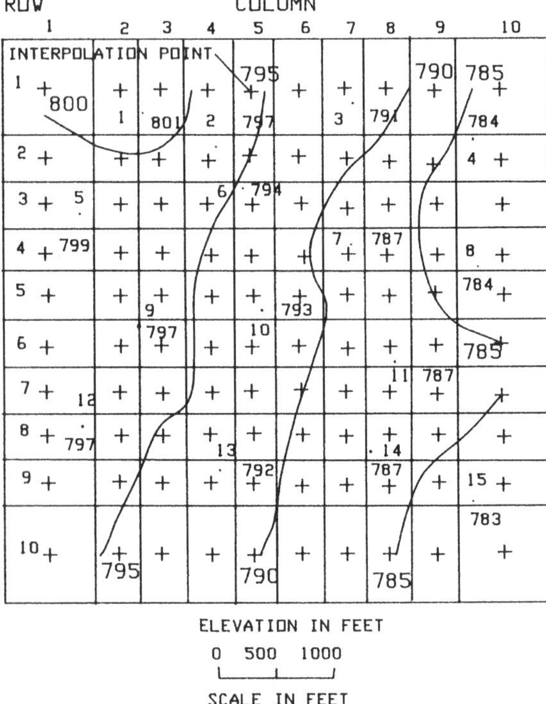

Figure 10. Scattered control data points and contour map for Exercise DM4.

Table 14. Scattered Control Point Data for Exercise DM4

Point Number	X-Coordinate (ft)	Y-Coordinate (ft)	Z Value (ft)
1	1850	1290	801
2	2475	1510	797
3	3700	1080	791
4	4680	1490	784
5	1400	2100	799
6	2910	1805	794
7	3670	2320	787
8	4720	2510	784
9	1800	3000	797
10	2830	2905	793
11	4070	3370	787
12	1310	3910	797
13	2490	4300	792
14	3840	4120	787
15	4700	4550	783

Table 15. Grid Spacing Data for Exercise DM4

Column	Grid Spacing (ft)	Row	Grid Spacing (ft)
1	800	1	800
2–9	400	2–9	400
10	800	10	800

dialog box. A Tabular Display spreadsheet and dialog box appears, prompting for information about the block number and the numerical format to be used in displaying values. Type "1" in the Block Number text box and select the "Decimal" Format option. Choose the "View" button to display XYZ values for the scattered triplet array in the spreadsheet box. After viewing the array using the scroll bars or arrow keys, choose the "OK" button to close the dialog box and return to the Manager.

Choose the "File" menu in the Manager Menu Bar, select the "Save" drop-down menu option, and select the "Scattered Triplet" and "Lower Origin" cascading options to save the entire scattered triplet array with a lower origin. A Specify File Dimensions data entry dialog box appears, prompting for information about point numbers, grid reference frame upper and lower Y-coordinates, and the name of the file in which the scattered triplet array will be stored. Type "1" in the First Point Number text box and "15" in the Last Point Number text box. Type a valid filename such as C:\DM\DM4S.DAT in the Save As text box. Type "1000" in the Reference Frame Upper Row Y-Coordinate text box, and type "5000" in the Reference Frame Lower Row Y-Coordinate text box. You have defined the control point reference frame so that the scattered triplet array stored in the computer memory with an upper left origin can be saved with a lower left origin. Choose the "Enter" button to store these entries, open the file, write the array, and close the file. Choose the "OK" button to close the dialog box and return to the Manager.

Choose the "Interpret" menu in the Manager Menu Bar, select the Surface drop-down menu option, and select the "Triangulation" cascading option. A Specify Grid Spacing Dimensions data entry dialog box appears, prompting for information about interpolation grid spacings and displaying the block number 1. Type "1" in the Block Column Data First Number and Last Number text boxes. Type "800" in the Block Column Data Grid spacing text box. Choose the "Enter Column" button to store these entries. Choose the "Next Column" button to clear text boxes and reset the focus so that spacings for other columns can be entered. Type "2" in the Block Column Data First Number text box and "9" in the Block Column Data Last Number text box. Type "400" in the Block Column Data Grid Spacing text box. Choose the "Enter Column" button to store these entries and the "Next Column" button to clear text boxes and reset the focus. Repeat this process until data for all three column grid spacing blocks have been entered. Choose the "Row" button to reset the focus on Row text boxes.

Type "1" in the Block Row Data First Number and Last Number text boxes. Type "800" in the Block Row Data Grid Spacing text box. Choose the "Enter Row button to store these entries. Choose the "Next Row" button to clear text boxes and reset the focus so that grid spacings for other rows can be entered. Type "2" in the Block Row Data First Number text box and "9" in the Block Row Data Last Number text box. Type "400" in the Block Row Data Grid Spacing text box. Choose the "Enter Column" button to store these entries and the "Next Row" button to clear text boxes and reset the focus. Repeat this process until data for all three row grid spacing blocks have been entered.

Type "1000" in the Upper Left Node X-Coordinate and Upper Left Node Y-Coordinate text boxes. Type "1" in the Left Column Number text box, "10" in the Right Column Number text box, "1" in the Upper Row Number text box, and "10" in the Lower Row Number text box. Notice that these dimensions are for the entire interpolation grid consisting of one or more zone blocks, whereas the previous dimensions were those for individual column and row blocks. Choose the "Enter Grid" button to store these entries. The interpolation gridded triplet array with an upper left origin and a variable grid spacing is now stored in the computer's memory and can be viewed and saved. Choose the "OK" button to close the dialog box.

A Specify Point Dimensions data entry dialog box appears, prompting for information about the scattered control data points. Type "1" in the First Point Number text box, and "15" in the Last Point Number text box. Choose the "Enter" button to store these entries, interpret between scattered control data points and the interpolation grid, and create the gridded XYZ triplet array. Choose the "OK" button to close the dialog box and return to the Manager.

Choose the "View" menu in the Manager Menu Bar, select the "Map" drop-down menu option, and select the "Contour" and "Screen" cascading options to view a contour map of the array. A Specify View Block Dimensions data entry dialog box appears, prompting for information about the viewport block dimensions and displaying 1 in the Block Number text box. Type "1" in the Block Upper Left Corner Column and Row Number text boxes. Type "10" in the Block Lower Right Corner Column and Row Number text boxes. Type "800" in the Block Upper Left Grid Spacing Column and Row text boxes. Type "FEET" in the X- and Y-Coordinate Unit and Z Value Unit text boxes. Choose the "Enter" button to store these entries. Choose the "OK" button to close the dialog box.

A Contour Map display picture and dialog box appears, prompting for information about the block number and the values of the contours to be viewed. The X- and Y-coordinate and Z value units are displayed in text boxes to the right of the picture box. Type "1" in the Block Number text box. The number 1 appears in the Contour Number text box. Type "785" in the Contour Value text box. Choose the "Enter" button to store these entries. The following information is displayed in the picture box: Block Upper Left Column = 1, Block Upper Left Row = 1, Block Lower Right Column = 10, Block Lower

Right Row = 10, Minimum Z Value = 7.832E+02, Minimum Z Value Column = 10, Minimum Z Value Row = 9, Maximum Z Value = 8.002E+02, Maximum Z Value Column = 2, and Maximum Z Value Row = 2. Choose the "Contour" button to clear the Contour Value text box so that the value for another contour can be entered. Repeat this process until contour values 790, 795, and 800 have been entered.

Choose the "Plot" button to view the contours and interpolation grid nodes. The information in the picture box disappears when contours appear. Choose the "Grid" button to display a cell border line grid on top of the contour map. Choose "Legend" to view the Z value of any contour. Point the cross-hair of the mouse at any contour, and the Z value for that contour will appear above the picture box when the left mouse button is depressed.

Choose the CR or XY Axes option and view column and row or XY coordinates at selected locations by depressing the left mouse button when the mouse pointer is at the desired location. Column and row or XY coordinates appear in text boxes to the right of the picture box. Compare the contour map with data in Table 12, thereby determining that the interpreted Z values are reasonable. Choose the "OK" button to close the dialog box and return to the Manager.

Any portion of the interpolation grid array or all of the interpolation grid array can be saved by specifying appropriate values for the file array grid dimensions. To save the entire interpolation grid array as a gridded triplet array with variable grid spacings and a lower left origin, choose the "File" menu in the Manager Menu Bar, select the "Save" drop-down menu option, and select the "Gridded Triplet" and "Lower Origin" cascading menu options. A Specify File Dimensions data entry dialog box appears, prompting for information about the file array dimensions. Type "1" in the Upper Left Corner Column and Row Number text boxes. Type "10" in the Lower Right Corner Column Number text box, and "10" in the Lower Right Corner Row Number text box. Type a valid name such as C:\DM\DM4GT.DAT for the file in which the array is to be saved in the Save As text box. Choose the "Enter" button to store these entries, open the file, writ the array, and close the file. Choose the "OK" button to close the dialog box and return to the Manager.

Choose the "Interpret" menu in the Manager Menu Bar, select the "Surface" drop-down menu option, and select the "Kriging" cascading option. A Specify Grid Spacing Dimensions data entry dialog appears prompting for information about grid spacings (interpolation grid) and displaying the block number 1. Type "1" in the Block Column Data First Number and Last Number text boxes. Type "800" in the Block Column Data Grid Spacing text box. Choose the "Enter Column" button to store these entries. Choose the "Next Column" button to clear text boxes and reset the focus so that spacings for other columns can be entered. Type "2" in the Block Column Data First Number text box and "9" in the Block Column Data Last Number text box. Type "400" in the Block Column Data Grid Spacing text box. Choose the "Enter" column button to store these entries and the Next Column button to clear text boxes and reset the focus. Repeat this process until data for all three column grid spacing blocks have been entered. Choose the "Row" button to reset the focus on Row text boxes.

Type "1" in the Block Row Data First Number and Last Number text boxes. Type "800" in the Block Row Data Grid Spacing text box. Choose the "Enter Row" button to store these entries. Choose the "Next Row" button to clear text boxes and reset the focus so that grid spacings for other rows can be entered. Type "2" in the Block Row Data First Number text box and "9" in the Block Row Data Last Number text box. Type "400" in the Block Row Data Grid Spacing text box. Choose the "Enter Column" button to store these entries and the "Next Row" button to clear text boxes and reset the focus. Repeat this process until data for all three row grid spacing blocks have been entered.

Type "1000" in the Upper Left Node X-Coordinate and Upper Left Node Y-Coordinate text boxes. Type "1" in the Left Column Number text box, "10" in the Right Column Number text box, "1" in the Upper Row Number text box, and "10" in the Lower Row Number text box. Notice that these dimensions are for the interpolation grid consisting of one or more zone blocks, whereas the previous dimensions were those for individual column and row blocks. Choose the "Enter Grid" button to store these entries. The gridded triplet array with an upper left origin and a variable grid spacing is now stored in the computer's memory and can be viewed and saved. Choose the "OK" button to close the dialog box.

A Specify Point Dimensions data entry dialog box appears, prompting for information about the control data points and the desired type of kriging results. Type "1" in the First Point Number text box and "15" in the Last Point Number text box. Select the "Z Value Kriging Array" Type option. Choose the "Enter" button to store these entries, to interpret between scattered control data points and the interpolation grid, and to create the gridded triplet XYZ array. Choose the "OK" button to close the dialog box and return to the Manager.

Choose the "View" menu in the Manager Menu Bar, select the "Map" drop-down menu option, and select the "Contour" and "Printer" cascading options to obtain a hard copy (takes several minutes) of the contour map of the gridded triplet krigged Z values. A Specify View Block Dimensions data entry dialog box appears, prompting for information about the view block dimensions. Type "1" in the Upper Left Corner Column and Row text boxes, and "10" in the Lower Right Corner Column and Row text boxes. Type "800" in the Upper Left Grid Spacing Column and Row text boxes. Choose the "Enter" button to store these entries. Choose the "OK" button to close the dialog box.

A Specify Printer Contour Map Dimensions data entry dialog box appears, prompting for information about the hard copy contour map dimensions and units. Type "1" in the Block Number text box. Type "FEET" in the Map X- and Y-Coordinate Unit and Z Value Unit text boxes. Choose the "Unit" button to store these entries. The numeral 1 appears in the Contour Number text box.

Type "786" in the Contour Value text box. Choose the "Contour" button to store this value. Choose the "Next Contour" button to clear the Contour Value text box so that values for another contour can be entered. Repeat this process until values for the 790, 795, and 798 contours have been entered. Choose the "Print" button to obtain a hard copy of the contour map with a cell border line grid and corner column and row and XY coordinates using *Windows* Print Manager. A contour value legend is printed below the map. Compare the contour map with data in Table 14, thereby determining that the interpreted Z values are reasonable. Choose the "OK" button to close the dialog box.

Any portion of the interpolation grid array or all of the interpolation grid array can be saved by specifying appropriate values for the file array grid dimensions. To save the entire interpolation grid array as a gridded triplet array with variable grid spacings and a lower left origin, choose the "File" menu in the Manager Menu Bar, select the "Save" drop-down menu option, and select the Gridded Triplet and Lower Origin cascading menu options. A Specify File Dimensions data entry dialog box appears, prompting for information about the file array dimensions. Type "1" in the Upper Left Corner Column and Row Number text boxes. Type "10" in the Lower Right Corner Column Number and Lower Right Corner Row Number text boxes. Type a valid name such as C:\DM\DM4GK.DAT for the file in which the array is to be saved in the Save As text box. Choose the "Enter" button to store these entries, open the file, write the array, and close the file. Choose the "OK" button to close the dialog box and return to the Manager.

Choose the "File" menu in the *DesignMod* Menu Bar and the "Exit" drop-down menu option. An Exit DesignMod message dialog box appears. Choose the "OK" button to exit *DesignMod*.

DM5 — Convert Units

This exercise helps you master File, Edit, and View menus and associated dialog boxes as you import a scattered triplet array, convert the units of the scattered triplet array, and edit and view a gridded triplet array. The scattered triplet array whose units are to be converted is that described in Exercise DM4 with 15 control points. The array is stored in file DM4S.DAT on the disk and contains X- and Y-coordinate and water level values. The gridded triplet array to be edited is that described in Exercise DM3 with 20 columns and 12 rows subdivided into 6 Z value zone blocks. The array upper left node X- and Y-coordinates are each 500. The array is stored in file DM3G.DAT on the disk and contains X- and Y-coordinate and hydraulic conductivity values.

Start *DesignMod* and choose the "File" menu in the Manager Menu Bar, select the "Load" drop-down menu option, and select the "Scattered Triplet and Lower Origin" cascading options. A Filename dialog box appears, prompting for information about the name of the file to be imported. Type C:\DM\DM4S.DAT or a similar specification in the Filename text box and choose the "OK" button, or choose the appropriate drive from the Drive List box and the appropriate directory from the Directory List box. Edit the Wildcard File List Extension text box to read *.DAT to limit the filenames displayed in the File List box. Choose the filename "DM4S.DAT" from the File List box and choose the "OK" button to close the dialog box.

A Specify File Dimensions data entry dialog appears, prompting for information about the scattered triplet array dimensions. Type "1" in the First Point Number text box, and "15" in the Last Point Number text box. Type "1000" in the Reference Frame Upper Row Y-Coordinate text box, and "5000" in the Reference Frame Lower Row Y-Coordinate text box. Choose the "Enter" button to store these entries. Choose the "OK" button to open the file, read the array, close the file, close the dialog box, and return to the Manager.

Assume that XYZ array values are to be converted from feet to meter units. Choose the "Edit" menu in the Manager Menu Bar, select the "Units" drop-down menu option, and select the "English To Metric" cascading option. A Specify Units dialog box appears, prompting for information about the conversion. Choose the "Length" dimension and the "ft-m" units. The conversion factor (.3048) appears in the Conversion Factor text box. Make a note of the conversion factor.

Choose the "Array" button. A Specify Conversion Dimensions dialog box appears. Select the "Triplet-S Array" Type option because you loaded a scattered triplet array. Type "1" in the Point Data First Number text box, and "15" in the Point Data Last Number text box. Type ".3048" in the X-, Y-, and Z-Conversion Factor text boxes. Choose the "Enter" button to store these entries and convert all array values. Choose the "OK" button to close the dialog box and return to the Specify Units dialog box. Choose the "OK" button in the Specify Units dialog box to return to the Manager.

Choose the "View" menu in the Manager Menu Bar; select the "Tabular" drop-down menu option; and select the "Triplet", "Scattered", and "Screen" cascading options to view the entire converted scattered triplet array. A Specify Tabular Display Dimensions data entry dialog box appears, prompting for information about the viewport block dimensions. Type "1" in the Block Number text box. Type "1" in the First Point Number text box, "15" in the Last Point Number text box, and "METERS" in the X- and Y-Coordinate Unit and Z Value Unit text boxes. Choose the "Enter" button to store these entries. Choose the "OK" button to close the dialog box.

A Tabular Display spreadsheet and dialog box appears, prompting for information about the block number and the numerical format to be used in displaying values. Type "1" in the Block Number text box and select the "Decimal" Format option. Choose the "View" button to display XYZ values for the scattered triplet array in the grid box. Compare the tabular display with Table 12, thereby confirming that all values have been properly converted. Choose the "OK" button to close the dialog box and return to the Manager.

To edit the gridded triplet array, choose the "File" menu in the Manager Menu Bar, select the "Load" drop-down menu option, and select the "Gridded Triplet" and "Lower Origin" cascading options. A Filename dialog box appears, prompting for information about the name of the file to be imported. Type C:\DM\DM3G.DAT or a similar specification in the Filename text box and choose the "Enter" button, or choose the appropriate drive from the Drive List box and the appropriate directory from the Directory List box. Edit the Wildcard File List Extension text box to read *.DAT to limit the filenames displayed in the File List box. Choose the filename "DM3G.DAT" from the File List box and the "Enter" button. Choose the "OK" button to close the dialog box.

A Specify File Dimensions data entry dialog box appears, prompting for information about the array dimensions. Type "1" in the Upper Left Corner Column and Row Number text boxes. Type "20" in the Lower Right Corner Column Number text box, and "12" in the Lower Right Corner Row Number text box. Choose the "Enter" button to store these entries. Choose the "OK" button to open the file, read the array, close the file, close the dialog box, and return to the Manager.

Choose the "Edit" menu in the Manager Menu Bar, select the Z or XYZ drop-down menu option, and select the Multiply cascading option to edit a portion of the array. A Specify Edit Block Dimensions data entry dialog box appears prompting for information about the block dimensions and operator value. The numeral 1 appears in the Block Number text box. Type "1" in the Upper Left Corner Column and Row Number text boxes. Type "7" in the Lower Right Corner Column Number text box and "5" in the Block Lower Right Corner Row Number text box. Select the "Gridded Triplet" and "Do Not Edit" options. Type "2" in the Value Or Operator Factor text box. Choose the "Enter" button to store these entries. Choose the "OK" button to multiply Z values within the block by 2, close the dialog box, and return to the Manager.

Choose the "View" menu in the Manager Menu Bar, select the "Tabular" drop-down menu option, and select the "Grid" and "Screen" cascading options to view only Z values, and not X- and Y-coordinates, within two viewport blocks. A Specify Tabular Display Dimensions data entry dialog box appears, prompting for information about the viewport block dimensions and the Z value unit. Type "1" in the Block Number text box. Choose the "Z Values Display" Type option. Type "1" in the Upper Left Corner Column and Row Number text boxes. Type "7" in the Lower Right Column Number text box, and "5" in the Lower Right Row Number text box. Type "GPD/SQ FT" in the Z Value Unit text box. Choose the "Enter" button to store these entries. Choose the "Next" button to clear text boxes and reset the focus so that values for block 2 can be entered. Type "2" in the Block Number text box. Type "1" in the Upper Left Corner Column and Row Number text boxes. Type "20" in the Lower Right Corner Column Number text box and "12" in the Lower Right

Corner Row Number text box. Choose the "Enter" button to store these entries. Choose the "OK" button to close the dialog box.

A Tabular Display spreadsheet and dialog box appears, prompting for information about the number of the block to be viewed and the type of numerical format to be used in displaying values and displaying the Z Value Unit. Type "1" in the Block Number text box and select the "Decimal" Format option. Choose the "View" button and Z values for block number 1 appear in the spreadsheet cells. Compare the Z values in the spreadsheet with those presented in Table 11 to verify that the original Z values in columns 1 to 7 and rows 1 to 5 have been multiplied by 2 as intended. Choose the "Next" button to view Z values in the entire edited array (block 2). Type "2" in the Block Number text box and select the "Decimal" Format option. Choose the "View" button to view Z values for block number 2. Compare the Z values in the spreadsheet with those presented in Table 12 to verify that Z values in only block 1 have been edited. Choose the "OK" button to close the dialog box and return to the Manager.

Choose the "File" menu in the *DesignMod* Menu Bar and the "Exit" drop-down menu option. An Exit DesignMod message dialog box appears. Choose the "OK" button to exit *DesignMod*.

DM6 — Edit Z Values

This exercise helps you master Edit, File, and View menus and associated dialog boxes while creating, editing, and viewing a grid array. The grid array has 18 columns and 11 rows as shown in Figure 11. The array is subdivided into 13 zone blocks with Z values as presented in Table 16 (file DM6G.DAT on the disk).

The Z values represent hydraulic conductivity values. Start *DesignMod*, choose the "Edit" menu in the Manager Menu Bar, select the "Z Or XYZ" drop-down menu option, and select the "New" cascading option. A Specify Edit Block Dimensions data entry dialog box appears prompting for information about the dimensions of each of the 13 blocks that comprise the grid array to be created. The numeral 1 appears in the Block Number text box. Type "1" in the Upper Left Corner Column and Row Number text boxes. Type "7" in the Lower Right Corner Column Number text box, and "3" in the Lower Right Corner Row Number text box. Select the "Grid Array" Type option. Type "500" in the Value Or Operator Factor text box. Choose the "Enter" button to store these entries. Choose the "Next" button to clear text boxes, increment the block number, and reset the focus so that values for another block can be entered. Type "8" in the Upper Left Corner Column text box, and "1" in the Upper Left Corner Row Number text box. Type "9" in the Lower Right Corner Column Number and "3" in the Lower Right Corner Row Number text box. Type "600" in the Value Or Operator Factor text box. Choose the "Enter" button to store these entries. Choose the "Next" button to clear text boxes,

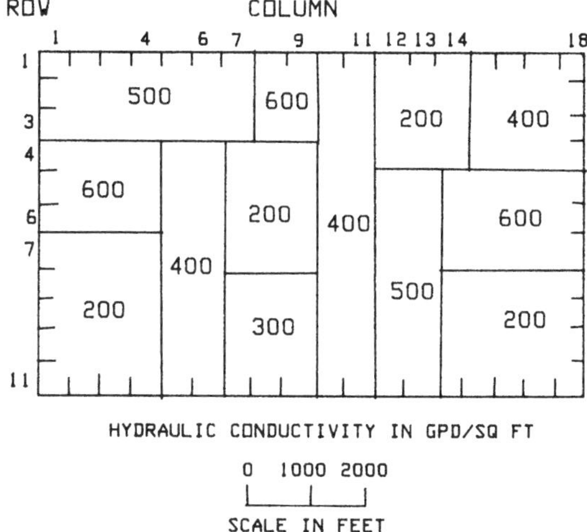

Figure 11. Z value zone blocks for Exercise DM6.

Table 16. Z Value Data for Exercise DM6

Column Number	Row Number	Z Value (gpd/sq ft)
1–7	1–3	500
8–9	1–3	600
10–11	1–11	400
12–14	1–4	200
15–18	1–4	400
1–4	4–6	600
5–6	4–11	400
1–4	7–11	200
7–9	4–7	200
7–9	8–11	300
12–13	5–11	500
14 18	5–7	600
14–18	8–11	200

increment the block number, and reset the focus so that values for another block can be entered. Repeat this process until values for all 13 blocks have been entered. Choose the "OK" button to close the dialog box and return to the Manager.

To save the entire array with a comma-separated value ASCII format, choose the "File" menu in the Manager Menu Bar, select the "Save" drop-down menu option, and select the "Grid" and "CSV" cascading options. A

Specify File Dimensions data entry dialog box appears, prompting for information about the dimensions of the array. Type "1" in the Upper Left Corner Column and Row Number text boxes. Type "18" in the Lower Right Corner Number text box. Type "11" in the Lower Right Corner Row Number text box. Type a valid name such as C:\DM\DM6G.DAT in the Save As text box. Choose the "Enter" button to store these entries. Choose the "OK" button to open a file, write the array, close the file, close the dialog box, and return to the Manager.

To replace the values in blocks having a Z value of 200 gpd/sq ft with a Z value of 280 gpd/sq ft, choose the "Edit" menu in the Manager Menu Bar, select the Z Or XYZ drop-down menu option, and select the "Search" and "Replace" cascading options. A Specify Edit Block Dimensions data entry dialog box appears, prompting for information about the dimensions of the block of columns and rows to be searched and the range of the search and displaying the block number 1. To search the entire array, type "1" in the Upper Left Corner Column and Row Number text boxes. Type "18" in the Lower Right Corner Column Number text box, and "11" in the Lower Right Corner Row Number text box. Select the "Grid Array" Type option. Type "280" (replacement value) in the Value Or Operator Factor text box. Type "300" in the Upper Range Value text box and "200" in the Lower Range Value text box. Choose the "Enter" button to store these entries. Choose the "OK" button to replace Z values that meet the search range criteria, close the dialog box, and return to the Manager.

Choose the "View" menu in the Manager Menu Bar, select the "Tabular" drop-down menu option, and select the "Grid" and "Screen" cascading options to view the entire array. A Specify Tabular Display Dimensions data entry dialog box appears, prompting for information about the dimensions of the viewport block to be viewed. Type "1" in the Block Number text box. Type "1" in the Upper Left Corner Column and Row Number text boxes. Type "18" in the Lower Right Corner Column Number text box and "11" in the Lower Right Corner Row Number text box. Type "GPD/SQ FT" in the Z Value Unit text box. Choose the "Enter" button to store these entries. Choose the "OK" button to close the dialog box.

A Tabular Display spreadsheet and dialog box appears, prompting for information about the viewport block number and the numerical format to be used in displaying values and displaying the Z Value Unit above the picture box. Type "1" in the Block Number text box and select the "Decimal" Format option. Choose the "View" button to view the entire contents of the array including the replace values. Compare the contents of the spreadsheet box with Table 16 to verify search and replace operations.

To copy part of the spreadsheet contents onto the Clipboard, position the mouse cursor in spreadsheet cell 1,1. Depress the left mouse button while moving the mouse cursor to spreadsheet cell 4,4, thereby highlighting a range of grid cells. Choose the "Clip" button to copy the contents of the highlighted spreadsheet cell range onto the *Windows* Clipboard. To view the contents of the Clipboard, choose *DesignMod*'s Control Menu box. The Tabular Display spreadsheet and dialog box disappears. Choose the "Switch To" option in the Control Menu box. A Task List box appears. Select the Program Manager in the Task List box. Choose the "Switch To" button. Select the Clipboard icon in the Main group. The Clipboard Viewer appears, displaying the contents of the Clipboard which should be the contents of the highlighted grid cell range. Choose the "File" menu in the Clipboard Viewer Menu Bar and the "Exit" drop-down menu option. Choose the Program Manager Control Menu box. Choose the "Switch To" option in the Control Menu box. Select "DesignMod Manager" in the Task List box. Choose the "Switch To" button. Press "Alt + F6" to recover the Tabular Display spreadsheet and dialog box. Choose the "OK" button to close the spreadsheet and dialog box and return to the Manager.

Choose the "File" menu in the *DesignMod* Menu Bar and the "Exit" drop-down menu option. An Exit DesignMod message dialog box appears. Choose the "OK" button to exit *DesignMod.*

DM7 — XY Graphs

This exercise helps you master the Edit, View, and File menus and associated dialog boxes while you create, view, and save arithmetic, semilog, and log-log XY arrays. The first part of the exercise involves the arithmetic array graph shown in Figure 12 with 11 points and the XY values presented in Table 17 (file DM7A.DAT on the disk).

The X values represent time and the Y values represent concentration. Start *DesignMod,* choose the "Edit" menu in the Manager Menu Bar, and select the "Doublet" drop-down menu option. A Specify Point Dimensions data entry dialog box appears, prompting for information about the array dimensions. Type "1" in the Point Number text box. Type "0" in the X-Coordinate text box, and "50" in the Y-Coordinate text box. Type "1" in the First Point Number text box and "11" in the Last Point Number text box. Choose the "Enter" button to store these entries. Choose the "Next" button to clear text boxes and reset the focus so that values for another point can be entered. Type "2" in the Point Number text box. Type "1" in the X-Coordinate text box and "70" in the Y-Coordinate text box. Choose the "Enter" button to store these entries. Choose the "Next" button to clear text boxes and reset the focus so that values for another point can be entered. Repeat this process until values for all 11 points have been entered. Choose the "OK" button to close the dialog box and return to the Manager.

To view array values, choose the "View" menu in the Manager Menu Bar, select the "Tabular" drop-down menu option, and select the "Doublet" and "Screen" cascading options. A Specify Tabular Display Dimensions data entry dialog box appears prompting for information about the dimensions of the viewport block. Type "1" in the Block Number text box. Type "1" in the First

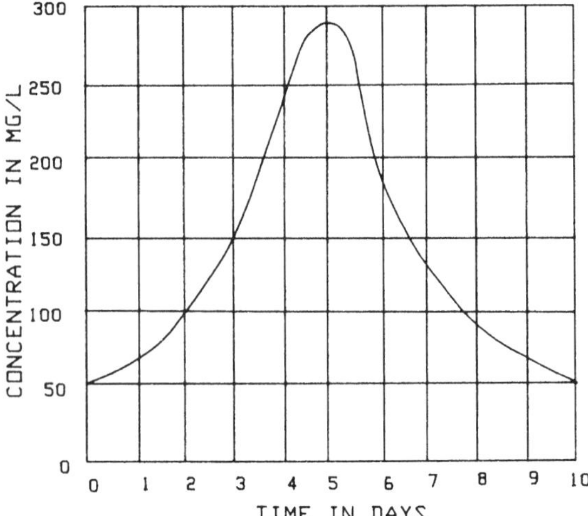

Figure 12.   Arithmetic graph for Exercise DM7.

Table 17. Arithmetic XY Data for Exercise DM7

Point Number	X-Coordinate Time (days)	Y-Coordinate Conc (mg/l)
1	0	50
2	1	70
3	2	100
4	3	150
5	4	240
6	5	280
7	6	180
8	7	130
9	8	90
10	9	70
11	10	50

Point Number text box and "11" in the Last Point Number text box. Type "TIME DAYS" in the X-Coordinate Unit text box and "CONC MG/L" in the Y-Coordinate Unit text box. Choose the "Enter" button to store these entries. Choose the "OK" button to close the dialog box.

A Tabular Display spreadsheet and dialog box appears, requesting information about the block number and the numerical format to be used in displaying values. Type "1" in the Block Number text box and select the "Decimal" Format option. Choose the "View "button to display the contents of the XY array. Use the scroll bars or the arrow keys to view the entire array. Choose the "OK" button to close the spreadsheet and dialog box and return to the Manager.

To save the entire array, choose the "File" menu in the Manager Menu Bar, select the "Save" drop-down menu option, and select the "Doublet" cascading option. A Specify File Dimensions data entry dialog box appears, prompting for information about the dimensions of the array. Type "1" in the First Point Number text box, and "11" in the Last Point Number text box. Type a valid name such as C:\DM\DM7A.DAT in the Save As text box. Choose the "Enter" button to store these entries. Choose the "OK" button to open the file, write the array, close the file, close the dialog box, and return to the Manager.

To view a graph of the array, choose the "View" menu in the Manager Menu Bar, select the "Graph" drop-down menu option, and select the "Arithmetic" and "Screen" cascading options. A Specify Screen Graph Dimensions data entry dialog box appears, prompting for information about the dimensions of the graph. Type "0" in the X-Axis Start Value text box, "10" in the X-Axis End Value text box, "0" in the Y-Axis Start Value text box, and "350" in the Y-Axis End Value text box. Type "1" in the Axis X-Grid Spacing text box and "50" in the Axis Y-Grid Spacing text box. Type "TIME DAYS" in the X-Axis Label And Unit text box and "CONC MG/L" in the Y-Axis Label And Unit text box. Choose the "Enter" button to store these entries. Choose the "OK" button to close the dialog box.

A Graph Display picture and dialog box appears, prompting for information about the graph features and displaying the X- and Y-axis labels and units. Choose the "Lower Origin" and the "Line Display" options. Choose the "Plot" button to display the graph and the "Grid" button to overlay the graph with a grid. Move the mouse pointer to a point on the graph and depress the left mouse button. The X- and Y-coordinates of the point are displayed to the lower right of the picture box. Choose the "OK" button to return to the Manager.

To obtain a hard copy of the graph, choose the "View" menu in the Manager Menu Bar, select the "Graph" drop-down menu option, and select the "Arithmetic" and "Printer" cascading options. A Specify Printer Graph Dimensions data entry dialog box appears prompting for information about the dimensions of the graph. Type "0" in the X-Axis Start Value text box, "10" in the X-Axis End Value text box, "0" in the Y-Axis Start Value text box, and "350" in the Y-Axis End Value text box. Type "TIME DAYS" in the X-Axis Label And Unit text box and "CONC MG/L" in the Y-Axis Label And Unit text box. Type "1" in the Axis X-Grid Spacing text box, and "50" in the Axis Y-Grid Spacing text box. Select the "Lower Origin", "Line Display", and "Grid Axes" options. Choose the "Enter" button to store these entries. Choose the "Print" button to obtain a hard copy of the graph. Choose the "OK" button to close the dialog box.

The second part of the exercise involves the semilogarithmic array graph shown in Figure 13 with 14 points and the XY values presented in Table 18 (file DM7S.DAT on the disk). The X values represent time and the Y values represent drawdown. Choose the "Edit" menu in the Manager Menu Bar and select the "Doublet" drop-down menu option. A Specify Point Dimensions data entry dialog box appears, prompting for information about the array dimensions. Type "1" in the Point Number text box. Type "15" in the X-Coordinate text box and "24.8" in the Y-Coordinate text box. Type "1" in the First Point Number text box and "14" in the Last Point Number text box. Choose the "Enter" button to store these entries. Choose the "Next" button to clear text boxes and reset the focus so that values for another point can be entered. Type "2" in the Point Number text box. Type "25" in the X-Coordinate text box and "25.5" in the Y-Coordinate text box. Choose the "Enter" button to store these entries. Choose the "Next" button to clear text boxes and reset the focus so that values for another point can be entered. Repeat this process until values for all 14 points have been entered. Choose the "OK" button to close the dialog box and return to the Manager.

To save the array, choose the "File" menu in the Manager Menu Bar, select the "Save" drop-down menu option, and select the "Doublet" cascading option. A Specify File Dimensions data entry dialog box appears, prompting for information about the dimensions of the array. Type "1" in the First Point Number text box, and "14" in the Last Point Number text box. Type a valid name such as C:\DM\DM7S.DAT in the Save As text box. Choose the "Enter" button to store these entries. Choose the "OK" button to open a file, write the array, close the file, close the dialog box, and return to the Manager.

To view the array graph, choose the "View" menu in the Manager Menu Bar, select the "Graph" drop-down menu option, and select the "Semilog" and "Screen" cascading options. A Specify Screen Graph Dimensions data entry dialog box appears prompting for information about the dimensions of the graph. Type "10" in the X-Axis Start Value text box, and "1000" in the X-Axis End Value text box. Type "24" in the Y-Axis Start Value text box and "32" in the Y-Axis End Value text box. Type "2" in the X-Axis Log Cycles text box and "2" in the Y-Axis Grid Spacing text box. Type "TIME MINUTES" in the X-Axis Label And Unit text box, and "DRAWDOWN FT" in the Y-Axis Label and Unit text box. Choose the "Enter" button to store these entries. Choose the "OK" button to close the dialog box.

A Graph Display picture and dialog box appears, prompting for information about the graph features and displaying the X- and Y-axis labels and units. Choose the Upper Origin and the Line Display options. Choose the "Plot" button to display the graph, and the "Grid" button to overlay the graph with a grid. Move the mouse pointer to a point on the graph and depress the left mouse button. The X- and Y-coordinates are displayed in text boxes to the right of the picture box. Choose the "OK" button to return to the Manager.

The third part of the exercise involves the logarithmic array graph shown in Figure 14 with 20 points and the XY values presented in Table 19 (file DM7L.DAT on the disk). The X values represent time and the Y values represent drawdown. Choose the "Edit" menu in the Manager Menu Bar, and

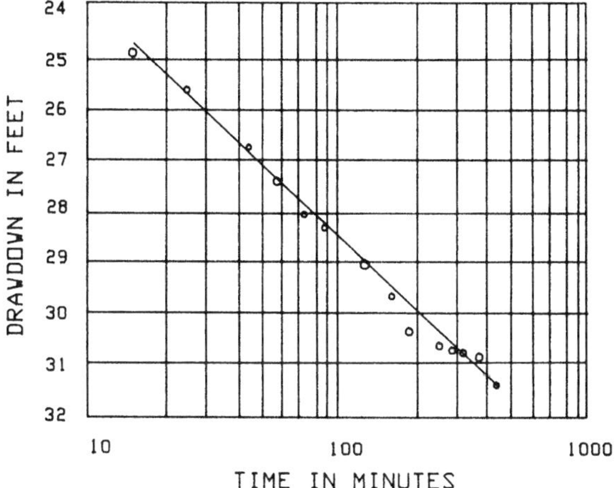

Figure 13.   Semilog XY graph for Exercise DM7.

Table 18. Semilog XY Data for Exercise DM7

Point Number	X-Coordinate Time (min)	Y-Coordinate Drawdown (ft)
1	15	24.8
2	25	25.5
3	45	26.6
4	60	27.3
5	76	28.0
6	90	28.2
7	132	29.0
8	166	29.5
9	195	30.3
10	256	30.5
11	282	30.6
12	314	30.7
13	360	30.8
14	430	31.5

select the "Doublet" drop-down menu option. A Specify Point Dimensions data entry dialog box appears, prompting for information about the array dimensions. Type "1" in the Point Number text box. Type "3" in the X-Coordinate text box, and "0.3" in the Y-Coordinate text box. Type "1" in the First Point Number text box and "20" in the Last Point Number text box. Choose the "Enter" button to store these entries. Choose the "Next" button to

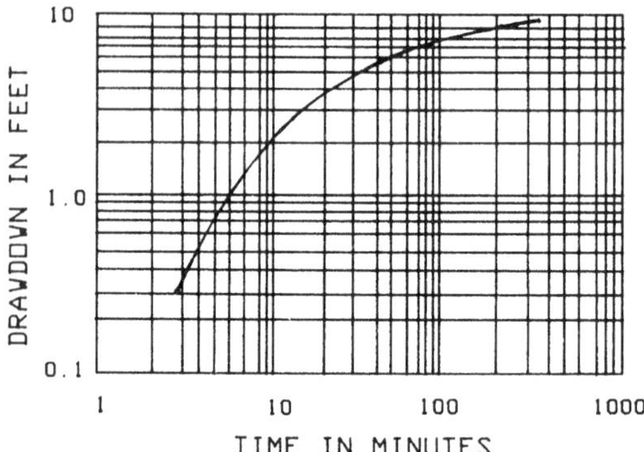

Figure 14.   Log XY graph for Exercise DM7.

clear text boxes and reset the focus so that values for another point can be entered. Type "2" in the Point Number text box. Type "5" in the X-Coordinate text box, and "0.7" in the Y-Coordinate text box. Choose the "Enter" button to store these entries. Choose the "Next" button to clear text boxes and reset the focus so that values for another point can be entered. Repeat this process until values for all 20 points have been entered. Choose the "OK" button to close the dialog box and return to the Manager.

To save the array, choose the "File" menu in the Manager Menu Bar, select the "Save" drop-down menu option, and select the "Doublet" cascading option. A Specify File Dimensions data entry dialog box appears, prompting for information about the dimensions of the array to be saved. Type "1" in the First Point Number text box, and "20" in the Last Point Number text box. Type a valid filename such as C:\DM\DM7L.DAT in the Save As text box. Choose the "Enter" button to store these entries. Choose the "OK" button to open a file, write the array, close the file, close the dialog box, and return to the Manager.

To print the array graph, choose the "View" menu in the Manager Menu Bar, select the "Graph" drop-down menu option, and select the "Logarithmic" and "Printer" cascading options. A Specify Printer Graph Dimensions data entry dialog box appears prompting for information about the dimensions of the graph. Type "1" in the X-Axis Start Value text box and "1000" in the X-Axis End Value text box. Type "0.1" in the Y-Axis Start Value text box and "10" in the Y-Axis End Value text box. Type "TIME MINUTES" in the X-Axis Label And Unit text box and "DRAWDOWN FT" in the Y-Axis Label And Unit text box. Type "3" in the X-Axis Log Cycles text box and "2" in the Y-Axis Log Cycles text box. Select the "Line Display" and "Grid Axes" options. Choose the "Enter" button to store these entries. Choose the "Print" button to obtain a hard copy of the graph. Choose the "OK" button to close the dialog box.

Table 19. Log XY Data for Exercise DM7

Point Number	X-Coordinate Time (min)	Y-Coordinate Drawdown (ft)
1	3	0.3
2	5	0.7
3	8	1.3
4	12	2.1
5	20	3.2
6	24	3.6
7	30	4.1
8	38	4.7
9	47	5.1
10	50	5.3
11	60	5.7
12	70	6.1
13	80	6.3
14	90	6.7
15	100	7.0
16	130	7.5
17	160	8.3
18	200	8.5
19	260	9.2
20	320	9.7

Choose the "File" menu in the *DesignMod* Menu Bar and the "Exit" drop-down menu option. An Exit DesignMod message dialog box appears. Choose the "OK" button to exit *DesignMod*.

DM8 — Mean Values

This exercise helps you master the Interpret menu and the Area, Mean, Sinks, and Time Steps drop-down menu options and associated dialog boxes. Part 1 of the exercise involves the calculation of the polygon area enclosed within 11 vertices. The X- and Y-coordinates of the vertices are presented in Table 20.

The X-coordinate represents time and the Y-coordinate represents discharge rate. Start *DesignMod*, choose the "Interpret" menu in the Manager Menu Bar, and select the "Area" drop-down menu option. A Specify Vertices Dimensions data entry dialog box appears, prompting for information about the vertice X- and Y-coordinates and displaying the vertice number 1. Type "11" in the Number Of Vertices text box. Type "0" in the X-Coordinate and the

**Table 20. Vertice Data for
Exercise DM8**

Vertex Number	X-Coordinate Time (min)	Y-Coordinate Discharge (gpm)
1	0	0
2	10	120
3	15	200
4	30	280
5	40	350
6	50	390
7	60	450
8	75	350
9	80	275
10	90	150
11	100	0

Y-Coordinate text boxes. Choose the "Enter" button to store these entries, and the "Next" button to increment the vertice number, clear text boxes, and reset the focus so that values for another vertice can be entered. Type "10" in the X-Coordinate text box and "120" in the Y-Coordinate text box. Choose the "Enter" button to store these entries and the "Next" button to increment the vertice number, clear text boxes, and reset the focus so that values for another vertice can be entered. Repeat this process until the values for all 11 vertices have been entered. Choose the "Area" button to calculate and display the area (26,488) within the vertices. Choose the "OK" button to close the dialog box and return to the Manager.

The second part of the exercise involves the calculation of the arithmetic, harmonic, and geometric means of the values presented in Table 21. Choose the "Interpret" menu in the Manager Menu Bar, select the "Mean" drop-down menu option, and select the "Arithmetic" cascading option. A Specify Mean Element Dimensions data entry dialog box appears, prompting for information about the values whose mean is to be calculated. The numeral 1 appears in the Element Number text box. Type "6" in the Number Of Elements text box and "18.4" in the Value text box. Choose the "Enter" button to store these entries and the "Next" button to clear text boxes and reset the focus so that additional values can be entered. Repeat this process until all six values are entered. Choose the "Mean" button to calculate and display the arithmetic mean of the values (18.33). Choose the "OK" button to close the dialog box and return to the Manager. Repeat the process just described, choosing the "Interpret" menu and selecting the "Mean" drop-down menu option and the "Harmonic" and "Geometric" cascading options in that order. The calculated harmonic and geometric mean values are 18.22 and 18.12, respectively.

Table 21. Mean Value Data for Exercise DM8

Value Number	Value
1	18.4
2	20.6
3	15.7
4	21.1
5	17.8
6	16.4

The third part of the exercise involves the calculation of the X- and Y-coordinates and discharge rate of a single (equivalent) sink simulating five individual sinks with the dimensions presented in Table 22. Choose the "Interpret" menu in the Manager Menu Bar and select the "Sinks" drop-down menu option. A Specify Sink Or Source Dimensions data entry dialog box appears, prompting for information about the dimensions of the sinks. The numeral 1 appears in the Sink Or Source Number text box. Type "5" in the Number Of Individual Sinks Or Sources text box. Type "1200" in the Individual Sink Or Source X-Coordinate text box. Type "1400" in the Individual Sink Or Source Y-Coordinate text box. type "200" in the Individual Sink Or Source Discharge text box. Choose the "Enter" button to store these entries and the "Next" button to increment the sink number, clear text boxes, and reset the focus so that values for another sink can be entered. Type "2400" in the Individual Sink Or Source X-Coordinate text box. Type "1200" in the Individual Sink Or Source Y-Coordinate text box. Type "200" in the Individual Sink Or Source Discharge text box. Choose the "Enter" button to store these entries and the "Next" button to increment the sink number, clear text boxes, and reset the focus so that values for another sink can be entered. Repeat this process until values for all five sinks have been entered. Choose the "Single" button to calculate and display the single sink X- and Y-coordinates (1760 feet, 1893 feet) and the discharge rate (300 gpm) of the single sink. Choose the "OK" button to close the dialog box and return to the Manager.

The fourth part of the exercise involves calculating a mean vertical hydraulic conductivity using the values presented in Table 23. Choose the "Interpret" menu in the Manager Menu Bar, select the "Mean" drop-down menu option, and select the "Vertical" cascading option. A Specify Mean Element Dimensions dialog box appears, prompting for information about the layer dimensions. The numeral 1 appears in the Element Number text box. Type "4" in the Number Of Elements text box, "0.01" in the Vertical Hydraulic Conductivity text box, and "3" in the Thickness text box. Choose the "Enter" button to store these values and the "Next" button to increment the element number, clear text boxes, and reset the focus so that additional values can be entered. Repeat this process until values for all four layers have been entered. Choose

Table 22. Sink Data for Exercise DM8

Sink Number	X-Coordinate (ft)	Y-Coordinate (ft)	Discharge Rate (gpm)
1	1200	1400	200
2	2400	1200	200
3	1500	2000	500
4	2200	2000	300
5	1700	2400	300

the "Mean" button to calculate and display the mean vertical hydraulic conductivity value (0.0363 gpd/sq ft). Choose the "OK" button to close the dialog box and return to the Manager.

The fifth part of the exercise involves calculating a mean horizontal hydraulic conductivity using the values presented in Table 24. Choose the "Interpret" menu in the Manager Menu Bar, select the "Mean" drop-down menu option, and select the "Horizontal" cascading option. A Specify Mean Element Dimensions dialog box appears, prompting for information about the layer dimensions. The numeral 1 appears in the Element Number text box. Type "4" in the Number Of Elements text box, "2000" in the Horizontal Hydraulic Conductivity text box, and "500" in the Thickness text box. Choose the "Enter" button to store these values and the "Next" button to increment the element number, clear text boxes, and reset the focus so that additional values can be entered. Repeat this process until values for all four layers have been entered. Choose the "Mean" button to calculate and display the mean horizontal hydraulic conductivity value (1730 gpd/sq ft). Choose the "OK" button to close the dialog box and return to the Manager.

The last part of the exercise involves calculating the initial time increment, flow critical time increment, and transport critical time increment for a transient transport model. Choose the "Interpret" menu in the Manager Menu Bar and select the "Time Step" drop-down menu option. A Specify Time Increments Dimensions data entry dialog box appears, prompting for information about the stress period and the representative model grid cell. Type "1" in the Stress Period Number text box. Type "50" in the Stress Period Number Of Time Steps text box. Type "365" (days) in the Stress Period Time At End Of Stress Period text box. Type "1.2" in the Stress Period Time Step Multiplier text box. Type "50" (feet) in the Representative Cell Grid Spacing text box. Type "10000" (sq ft/day) in the Representative Cell Transmissivity text box. Type "0.25" in the Representative Cell Storativity text box. Type "0.75" (ft/day) in the Representative Cell Velocity text box. Choose the "Enter" button to calculate and display the Initial Time Increment (0.008 day), Flow Critical Time Increment (0.0156 day), and Transport Critical Time Increment (66.67

**Table 23. Vertical Hydraulic Conductivity
Data for Exercise DM8**

Element Number	Vertical Hydraulic Conductivity (gpd/sq ft)	Layer Thickness (ft)
1	0.01	3
2	0.12	5
3	0.07	10
4	0.03	2

**Table 24. Horizontal Hydraulic Conductivity
Data for Exercise DM8**

Element Number	Horizontal Hydraulic Conductivity (gpd/sq ft)	Layer Width (ft)
1	2000	500
2	1500	200
3	1600	1000
4	1800	600

days). Choose the "OK" button to close the dialog box and return to the Manager.

Choose the "File" menu in the DesignMod Menu Bar and the "Exit" drop-down menu option. An Exit DesignMod message dialog box appears. Choose the "OK" button to exit *DesignMod.*

DM9 — File Catalog

This exercise helps you master the File menu and the drop-down Catalog menu option while you view and print a catalog of *DesignMod*'s exercise files. Choose the "File" menu in the Manager Menu Bar and select the "Catalog" drop-down menu option. A File Catalog menu and text box appears. Choose the "File" menu in the File Catalog Menu Bar and the "Load" drop-down menu option. A Specify Filename dialog box appears, prompting for the name of the file in which the contents of the file catalog are stored. The catalog is stored in file DM9.TXT on the disk. Type C:\DM\DM9.TXT or a similar specification in the Filename text box and choose the "OK" button, or choose the appropriate drive from the Drive List box and the appropriate directory from the Directory List box. Edit the Wildcard File List Extension text box to read *.DAT to limit the filenames displayed in the File List box. Choose the filename DM9.TXT. Choose the "OK" button to load the catalog and display it in the Existing File Catalog text box and close the Existing Filename dialog box.

View the catalog by using the scroll bars or the arrow keys. Choose the file menu in the Catalog Menu Bar and the "Exit" drop-down menu option to close the Existing File Catalog menu and text box and return to the Manager.

Choose the "File" menu in the *DesignMod* Menu Bar and the "Exit" drop-down menu option. An Exit DesignMod message dialog box appears. Choose the "OK" button to exit *DesignMod*.

REFERENCES

Anand, V. B., *Computer Graphics and Geometric Modeling for Engineers*, John Wiley & Sons, New York, 1993, 407 pp.

Anderson, M. P. and Woessner, W. W., *Applied Groundwater Modeling Simulation of Flow and Advective Transport*, Academic Press, New York, 1992, 381 pp.

Appleman, D., *Visual Basic Programmer's Guide to the Windows API*, Ziff-Davis Press, Emeryville, CA, 1993, 1020 pp.

Arnson, R., Gemmell, C., and Henderson, H., *MS-DOS QBasic Programmer's Reference*, Microsoft Press, Redmond, WA, 1991, 619 pp.

Arnson, R., Rosen, D., Waite, M., and Zuck, J., *Visual Basic How-To*, Waite Group Press, Corte Madera, CA, 1992, 546 pp.

Bonner, P., *Visual Basic Utilities*, Ziff-Davis Press, Emeryville, CA, 1993.

Boutwell, S. H., Brown, S. M., Roberts, B. R., and Atwood, D. F., *Modeling Remedial Actions at Uncontrolled Hazardous Waste Sites*, Noyes Publications, Park Ridge, NJ, 1986.

Clark, D., Microcomputer Programs for Groundwater Studies, *Developments In Water Science*, Vol. 30, Elsevier, New York, 1987.

Committee on Ground Water Modeling Assessment, *Ground Water Models — Scientific and Regulatory Applications*, National Academy of Sciences, Washington, D.C., 1990, 303 pp.

Cornell, G., *Visual Basic for Windows Inside and Out*, Osborne/McGraw-Hill, New York, 1992, 648 pp.

Craig, J. C., *Visual Basic Workshop*, Microsoft Press, Redmond, WA, 1991, 302 pp.

Davis, J. C., *Statistics and Data Analysis in Geology*, 2nd ed., John Wiley & Sons, New York, 1986, 646 pp.

Domenico, P. A. and Schwartz, F. W., *Physical and Chemical Hydrogeology*, John Wiley & Sons, New York, 1990, 824 pp.

Entsminger, G., *Secrets of the Visual Basic for Windows Masters*. Sams Publishing, Carmel, IN, 1992, 600 pp.

Environmental Systems Research Institute (ESRI), *Understanding GIS: The ARC/INFO Method*, John Wiley & Sons, New York, 1993.

Franke, O. L., Reilly, T. E., and Bennett, G. D., Definition of boundary and initial conditions in the analysis of saturated ground-water flow systems — an introduction, in *Techniques of Water-Resources Investigations*, Book 3, U.S. Geological Survey, 1987, chpt. B5.

Franke, O. L. and Reilly, T. E., The Effects of Boundary Conditions on the Steady-State Response of Three Hypothetical Ground-Water Systems — Results and Implications of Numerical Experiments, Water Supply Paper 2315, U.S. Geological Survey, 1987.

Glass, J. P., Groundwater modeling at the Seymour recycling superfund site, in *Solving Ground Water Problems with Models. Proceedings of the Fourth International Conference on the Use of Models to Analyze and Find Working Solutions to Ground Water Problems*, National Ground Water Association, Columbus, OH, 1989, pp. 1059-1081.

Goode, D. J. and Konikow, L. F., Modification of a method-of-characteristics solute-transport model to incorporate decay and equilibrium-controlled sorption or ion exchange, Water Resources Investigations Report 89-4030, U.S. Geological Survey, Denver, CO, 1989.

Gurewich, N. and Gurewich, O., *Teach Yourself Visual Basic 3.0 in 21 Days,* Sams Publishing, Carmel, IN, 1993, 967 pp.

Halvorson, M. and Rygmyr, D., *Running QBASIC,* Microsoft Press, Redmond, WA, 1991, 480 pp.

Hergert, D. A., *Visual Basic Programming,* Bantam Books, New York, 1991, 472 pp.

Holzner, S., *Visual Basic,* Brady Publishing, New York, 1991, 449 pp.

Hsieh, P. A. and Freckleton, J. R., Documentation of a computer program to simulate horizontal-flow barriers using the Modular Three-Dimensional Ground-Water Flow Model, Open File Report 92-477, U.S. Geological Survey, Denver, CO, 1992.

Kinzelbach, W., *Groundwater Modeling — An Introduction with Sample Programs in BASIC,* Elsevier, New York, 1986, 334 pp.

Konikow, L. F. and Bredehaeft, J. D., Computer model of two-dimensional solute transport and dispersion in ground water, in *U.S. Geological Survey Techniques of Water-Resources Investigations,* Book 7, U.S. Geological Survey, Denver, CO, 1978, chpt. C2.

Maxey, G. B., Hydrostratigraphic units, *J. Hydrology,* Vol. 2, 124-129, 1964.

Microsoft Programmer's Reference Library, *Windows 3.1 Programmer's Reference, Vol. 2, Functions,* Microsoft Press, Redmond, WA, 1992, 1001 pp.

Microsoft Programming Series, *The Windows Interface: An Application Design Guide,* Microsoft Press, Redmond, WA, 1992, 228 pp.

Microsoft Programming Series, *Windows Guide to Programming,* Microsoft Press, Redmond, WA, 1992.

Microsoft Visual Basic Programmer's Guide, Microsoft Press, Redmond, WA, 1993, 711 pp.

Microsoft Visual Basic Language Reference, Microsoft Press, Redmond, WA, 1993, 673 pp.

Microsoft Professional Features Book 1, Microsoft Press, Redmond, WA, 1993.

Microsoft Professional Features Book 2, Microsoft Press, Redmond, WA, 1993.

Microsoft Windows User's Guide, Microsoft Press, Redmond, WA, 1992.

Murray, W. H. and Pappas, C. H., *Using Visual Basic Writing Windows Applications,* Addison-Wesley Publishing, Reading, MA, 1992, 420 pp.

Orvis, W. J., *Do It Yourself Visual Basic,* Sams Publishing, Carmel, IN, 1992, 669 pp.

Poole, L., Borchers, M., and Koessel, K., *Some Common BASIC Programs,* Osborne/McGraw-Hill, New York, 1981, 193 pp.

Ribar, L. J., *FORTRAN Programming for Windows,* Osborne/McGraw-Hill, New York, 238 pp.

Rifai, H. S., Hendricks, L. A., Kilborn, K., and Bedient, P. B., A geographical information system (GIS) user interface for delineating wellhead protection areas, *Ground Water,* 31, (3), pp. 480-488, 1993.

Rumbaugh, J. O., Increasing the efficiency and accuracy of applied modeling using a database approach, in *Solving Ground Water Problems with Models. Proceedings of the Fourth International Conference on the Use of Models to Analyze and Find Working Solutions to Ground Water Problems,* National Ground Water Association, Columbus, OH, 1989, pp. 689-690.

Seaber, P. R., Hydrostratigraphics units, in *Hydrogeology,* Back, W. Rosenshein, J. S. and Seaber, P. R., Eds., *The Geology of North America,* Geological Society of America, 1988, 9-14.

Thomas, Z., Arnson, R., and Waite, M., *Visual Basic How-To 2E: The Definitive VB 3.0 Problem Solver,* Waite Group Press, Corte Madera, CA, 1993.

Trescott, P. C., Pinder, G. F. and Larson, S. P., Finite-difference model for aquifer simulation in two dimensions with results of numerical experiments, *Techniques of Water-Resources Investigations,* Book 7, U.S. Geological Survey, Denver, CO, 1976, chpt. C1.

Ward, D. S., Buss, D. R., Mercer, J. W., and Hughes, S. S., Evaluation of a groundwater corrective action at the Chem-Dyne hazardous waste site using a telescopic mesh refinement modeling approach, *Water Resources Research,* 23, (4), 603-617, 1987.

Walton, W. C., *Principles of Groundwater Engineering,* Lewis Publishers, Chelsea, MI, 1991, 546 pp.

Walton, W. C., *Groundwater Modeling Utilities,* Lewis Publishers, Chelsea, MI, 1992, 640 pp.

Weinman, D. G. and Kurshan, B. L., *IBM PC BASIC for Scientists and Engineers,* Reston Publishing, Reston, VA, 1985, 344 pp.

Young, M. J., *Visual Basic — Game Programming for Windows,* Microsoft Press, Redmond, WA, 1992, 513 pp.

3-D Widgets/2 Programmer's Guide, Sheridan Software Systems, Melville, NY, 1991.

3-D Widgets/3 Programmer's Guide, Sheridan Software Systems, Melville, NY, 1991.

APPENDIX A: EXAMPLE *VISUAL BASIC* *DESIGNMOD* SOURCE CODE

The following form description and *Visual Basic* code is for *DesignMod's* form DR7.FRM. This form enables the user to draw a box in the Design window with keyboard data entry.

FORM DESCRIPTION

VERSION 2.00

```
Begin Form DR7
   BackColor = &H00C0C0C0&
   BorderStyle = 3 'Fixed Double
   Caption = "Specify Box Dimensions"
   Height = 6240
   Left = 2175
   LinkMode = 1 'Source
   LinkTopic = "Form8"
   MaxButton = 0 'False
   MinButton = 0 'False
   ScaleHeight = 5835
   ScaleWidth = 5040
   Top = 525
   Width = 5160
   Begin SSCommand DR7Cmd4
      BevelWldth = 3
      Cation = "&Cancel"
      Font3D = 3 'Inset w/light shading
      FontBold = -1 'True
      FontItalic = 0 'False
      FontName = "MS Sans Serif"
      FontSize = 9.75
      FontStrikethru = 0 'False
```

```
      FontUnderline = 0 'False
      ForeColor = &H00000000&
      Height = 495
      Left = 2760
      TabIndex = 9
      Top = 5040
      Width = 975
End
Begin SSFrame DR7Fra1
   Alignment = 2 'Center
   Caption = "Coordinate Type"
   Font3D = 1 'Raised w/light shading
   ForeColor = &H00000000&
   Height = 855
   Left = 1080
   ShadowColor = 1 'Black
   TabIndex = 19
   Top = 240
   Width = 2415
   Begin SSOption DR7Opt2
      Caption = "Site"
      Font3D = 3 'Inset w/light shading
      ForeColor = &H00000000&
      Height = 255
      Left = 1440
      TabIndex = 1
      Top = 360
      Width = 735
   End
   Begin SSOption DR7Opt1
      Caption = "Screen"
      Font3D = 3 'Inset w/light shading
      ForeColor = &H00000000&
      Height = 255
      Left = 240
      TabIndex = 0
      Top = 360
      Width = 975
```

```
      End
   End
   Begin SSCommand DR7Cmd5
      BevelWidth = 3
      Caption = "&Help"
      Font3D = 3 'Inset w/light shading
      FontBold = -1 'True
      FontItalic = 0 'False
      FontName = "MS Sans Serif"
      FontSize = 9.75
      FontStrikethru = 0 'False
      FontUnderline = 0 'False
      ForeColor = &H00000000&
      Height = 495
      Left = 3720
      TabIndex = 10
      Top = 5040
      Width = 735
   End
   Begin SSCommand DR7Cmd3
      BevelWidth = 3
      Caption = "&OK"
      Font3D = 3 'Inset w/light shading
      FontBold = -1 'True
      FontItalic = 0 'False
      FontName = "MS Sans Serif"
      FontSize = 9.75
      FontStrikethru = 0 'False
      FontUnderline = 0 'False
      ForeColor = &H00000000&
      Height = 495
      Left = 1920
      TabIndex = 8
      Top = 5040
      Width = 855
   End
   Begin SSCommand DR7Cmd2
      BevelWidth = 3
```

```
      Caption = "&Next"
      Font3D = 3 'Inset w/light shading
      FontBold = -1 'True
      FontItalic = 0 'False
      FontName = "MS Sans Serif"
      FontSize = 9.75
      FontStrikethru = 0 'False
      FontUnderline = 0 'False
      ForeColor = &H00000000&
      Height = 495
      Left = 1200
      TabIndex = 7
      Top = 5040
      Width = 735
   End
   Begin SSCommand DR7Cmd1
      Bevelwidth = 3
      Caption = "&Draw"
      Font3D = 3 'Inset w/light shading
      FontBold = -1 'True
      FontItalic = 0 'False
      FontName = "MS Sans Serif"
      FontSize = 9.75
      FontStrikethru = 0 'False
      FontUnderline = 0 'False
      ForeColor = &H00000000&
      Height = 495
      Left = 480
      TabIndex 6
      Top = 5040
      Width = 735
   End
   Begin SSPanel DR7Pan7
      Alignment = 1 'Left Justify - MIDDLE
      BackColor = &H00C0C0C0&
      BevelOuter = 0 'None
      Font3D = 3 'Inset w/light shading
      ForeColor = &H00000000&
```

```
      Height = 495
      Left = 360
      TabIndex = 17
      Top = 3960
      Width = 2175
   End
   Begin SSPanel DR7Pan8
      Alignment = 1 'Left Justify - MIDDLE
      BackColor = &H00C0C0C0&
      BevelInner = 1 'Inset
      BevelOuter = 1 'Inset
      Font3D = 3 'Inset w/light shading
      ForeColor = &H00000000&
      Height = 495
      Left = 2880
      TabIndex = 18
      Top = 3960
      Width = 1695
      Begin TextBox DR7Text4
         Height = 285
         Left = 120
         TabIndex = 5
         Top = 120
         Width = 1455
      End
   End
   Begin SSPanel DR7Pan5
      Alignment = 1 'Left Justify - MIDDLE
      BackColor = &H00C0C0C0&
      BevelOuter = 0 'None
      Font3D = 3 'Inset w/light shading
      ForeColor = &H00000000&
      Height = 495
      Left = 360
      TabIndex = 15
      Top = 3120
      Width = 2175
   End
```

```
Begin SSPanel DR7Pan6
   Alignment = 1 'Left Justify - MIDDLE
   BackColor = &H00C0C0C0&
   BevelInner = 1 'Inset
   BevelOuter = 1 'Inset
   Font3D = 3 'Inset w/light shading
   ForeColor = &H00000000&
   Height = 495
   Left = 2880
   TabIndex = 16
   Top = 3120
   Width = 1695
   Begin TextBox DR7Text3
      Height = 285
      Left = 120
      TabIndex = 4
      Top = 120
      Width = 1455
   End
End
Begin SSPanel DR7Pan3
   Alignment = 1 'Left Justify - MIDDLE
   BackColor = &H00C0C0C0&
   BevelOuter = 0 'None
   Font3D = 3 'Inset w/light shading
   ForeColor = &H00000000&
   Height = 495
   Left = 360
   TabIndex = 13
   Top = 2280
   Width = 2175
End
Begin SSPanel DR7Pan4
   Alignment = 1 'Left Justify - MIDDLE
   BackColor = &H00C0C0C0&
   BevelInner = 1 'Inset
   BevelOuter = 1 'Inset
   Font3D = 3 'Inset w/light shading
```

```
      ForeColor = &H00000000&
      Height = 495
      Left = 2880
      TabIndex = 14
      Top = 2280
      Width = 1695
      Begin TextBox DR7Text2
         Height = 285
         Left = 120
         TabIndex = 3
         Top = 120
         Width = 1455
      End
   End
   Begin SSPanel DR7Pan1
      Alignment = 1 'Left Justify - MIDDLE
      BackColor = &H00C0C0C0&
      BevelOuter = 0 'None
      Font3D = 3 'Inset w/light shading
      ForeColor = &H00000000&
      Height = 495
      Left = 360
      TabIndex = 11
      Top = 1440
      Width = 2175
   End
   Begin SSPanel DR7Pan2
      Alignment = 1 'Left Justify - MIDDLE
      BackColor = &H00C0C0C0&
      BevelInner = 1 'Inset
      BevelOuter = 1 'Inset
      Font3D = 3 'Inset w/light shading
      ForeColor = &H00000000&
      Height = 495
      Left = 2880
      TabIndex = 12
      Top = 1440
      Width = 1695
```

```
    Begin TextBox DR7Text1
      Height = 285
      Left = 120
      TabIndex = 2
      Top = 120
      Width = 1455
    End
  End
End
```

CODE

```
DefSng A-Z

Sub DR7Cmd1_Click ()
  NL$ = Chr$(13) + Chr$(10)
  DR7Cmd2.Enabled = -1
  DR7Cmd3.Enabled = -1
  X1S = Val(DR7Text1.text)
  Y1S = Val(DR7Text2.text)
  X2S = Val(DR7Text3.text)
  Y2S = Val(DR7Text4.text)
  Screen.MousePointer = 11
  On Error GoTo CalculationError
  If (Flag8 = 1) Then
    X1 = X1S
    Y1 = Y1S
    X2 = X2S
    Y2 = Y2S
  End If
  If (Flag8 = 2 And Flag44 = 1) Then
    X1 = X1S
    Y1 = Y1S
    X2 = X2S
    Y2 = Y2S
  End If
  If (Flag8 = 2 And Flag44 = 2) Then
  'Transform site to model grid coordinates
```

```
  If (Flag40 = 1) Then
     YDifM = LRYC(K) − ULYC(K)
     X1S1 = (X1S − XOFF) * Cos(ROTANGR) + (Y1S − YOFF) *
Sin(ROTANGR)
     Y1S1 = (Y1S − YOFF) * Cos(ROTANGR) − (X1S − XOFF) *
Sin(ROTANGR)
     X2S1 = (X2S − XOFF) * Cos(ROTANGR) + (Y2S − YOFF) *
Sin(ROTANGR)
     Y2S1 = (Y2S − YOFF) * Cos(ROTANGR) − (X2S − XOFF) *
Sin(ROTANGR)
     X1 = Abs(X1S1)
     Y1 = Abs(YDifM − Y1S1)
     X2 = Abs(X2S1)
     Y2 = Abs(YDifM − Y2S1)
  End If
  If (Flag40 = 2) Then
     YDifM = LRYC(K) − ULYC(K)
     X1S1 = (X1S − XOFF) * Cos(ROTANGR) − (Y1S − YOFF) *
Sin(ROTANGR)
     Y1S1 = (Y1S − YOFF) * Cos(ROTANGR) + (X1S − XOFF) *
Sin(ROTANGR)
     X2S1 = (X2S − XOFF) * Cos(ROTANGR) + (Y2S − YOFF) *
Sin(ROTANGR)
     Y2S1 = (Y2S − YOFF) * Cos(ROTANGR) + (X2S − XOFF) *
Sin(ROTANGR)
     X1 = Abs(X1S1)
     Y1 = Abs(YDifM − Y1S1)
     X2 = Abs(X2S1)
     Y2 = Abs(YDifM − Y2S1)
  End If
End If
DR15.DR15Pict1.DrawWidth = Flag 3
DR15.DR15Pict1.DrawStyle = Flag1
DR15.DR15Pict1.FillColor = QBColor(Flag2)
DR15.DR15Pict1.FillStyle = Flag4
DR15.DR15Pict1.Line (X1, Y1)-(X2, Y2), QBColor(Flag2), B
Screen.MousePointer = 0
Exit Sub
```

```
CalculationError:
  Screen.MousePointer = 0
  Msg$ = "Calculation error" + Str$(Err) + "." + NL$
  Msg$ = Msg$ + "Try again with new data." + NL$
  Title$ = "Error Message"
  MsgBox Msg$, 48, Title$
  Exit Sub
End Sub

Sub DR7Cmd2_Click ()
  DR7Cmd2.Enabled = 0
  DR7Text1.text=""
  DR7Text2.text=""
  DR7Text3.text=""
  DR7Text4.text=""
  DR7Text1.SetFocus
End Sub

Sub DR7Cmd3_Click ()
  Flag1 = 0
  Flag2 = 0
  Flag3 = 1
  Flag4 = 1
  Unload DR7
End Sub

Sub DR7Cmd4_Click ()
  Flag1 = 0
  Flag2 = 0
  Flag3 = 1
  Flag4 = 1
  Flag44 = 0
  Unload DR7
End Sub

Sub DR7Cmd5_Click ()
  NL$ = Chr$(13) + Chr$(10)
```

```vb
    Msg$ = "Screen refers to coordinates above" + NL$
    Msg$ = Msg$ + "and left of picture box. Site refers" + NL$
    Msg$ = Msg$ + "to site base map coordinates within" + NL$
    Msg$ = Msg$ + "cross section or plan view boundar-" + NL$
    Msg$ = Msg$ + "ies. Box with previously selected" + NL$
    Msg$ = Msg$ + "color, style, and width is plotted" + NL$
    Msg$ = Msg$ + "with DRAW. NEXT clears text boxes so" + NL$
    Msg$ = Msg$ + "that another box can be plotted." + NL$
    Msg$ = Msg$ + "OK closes dialog box. CANCEL ends" + NL$
    Msg$ = Msg$ + "dialog before completion and" + NL$
    Msg$ = Msg$ + "closes dialog box." + NL$
    Title$ = "Help Message"
    MsgBoxMsg$, 64, Title$
End Sub


Sub DR7Opt1_Click (Value As Integer)
  If (Flag8 = 2) Then
    Flag44 = 1
    DR7Pan1.Caption = "Upper Left Corner Screen X-Coordinate:"
    DR7Pan3.Caption = "Upper Left Corner Screen Y-Coordinate:"
    DR7Pan5.Caption = "Lower Right Corner Screen X-Coordinate:"
    DR7Pan7.Caption = "Lower Right Corner Screen Y-Coordinate:"
  End If
End Sub


Sub DR7Opt2_Click (Value As Integer)
  If (Flag8 = 2) Then
    Flag44 = 2
    DR7Pan1.Caption = "Upper Left Corner Site X-Coordinate:"
    DR7Pan3.Caption = "Upper Left Corner Site X-Coordinate:"
    DR7Pan5.Caption = "Lower Right Corner Site Y-Coordinate:"
    DR7Pan7.Caption = "Lower Right Corner Site Y-Coordinate:"
  End If
End Sub


Sub Form_Load ()
  Top = Screen.Height/2 - Height/2
```

```
   Left = Screen.Width/2 - Width/2
  If (Flag8 = 1) Then
    DR7FRA1.Visible = 0
    DR7Opt1.Visible = 0
    DR7Opt2.Visible = 0
    DR7Pan1.Caption = "Upper Left Corner Screen X-Coordinate:"
    DR7Pan3.Caption = "Upper Left Corner Screen Z-Coordinate:"
    DR7Pan5.Caption = "Lower Right Corner Screen X-Coordinate:"
    DR7Pan7.Caption = "Lower Right Corner Screen Z-Coordinate:"
  End If
  DR7Cmd2.Enabled = 0
  DR7Cmd3.Enabled = 0
End Sub
```

APPENDIX B:
SAMPLE SCREENS

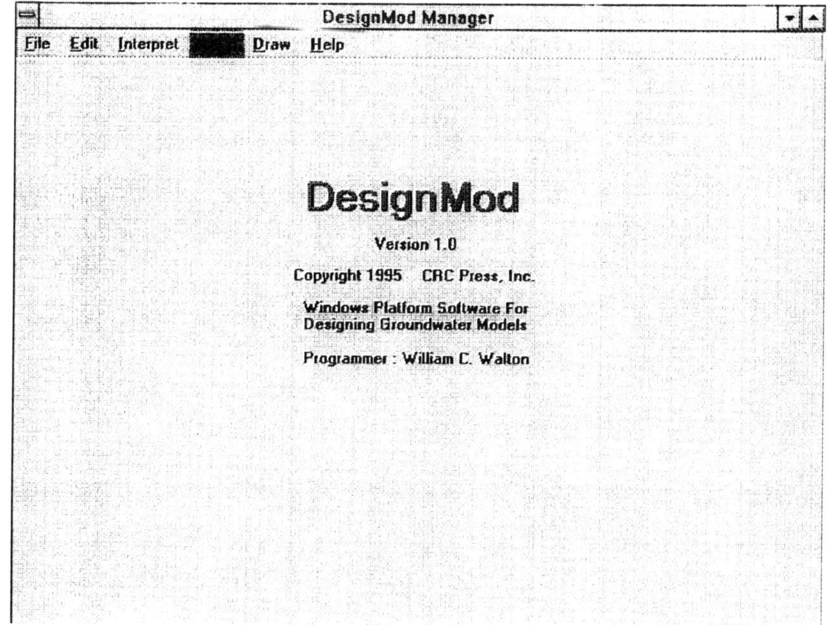

Figure B1.   Manager window at start up.

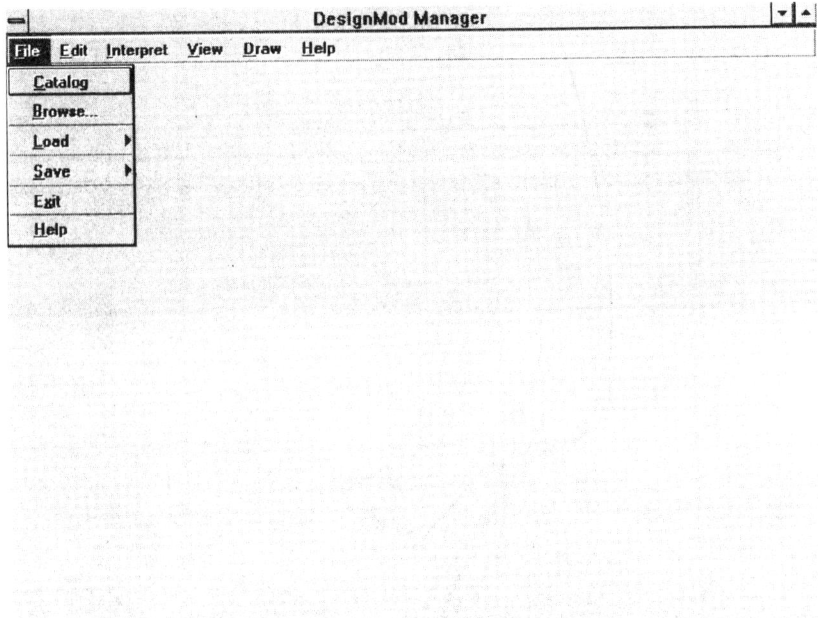

Figure B2.   Manager window with drop down menus.

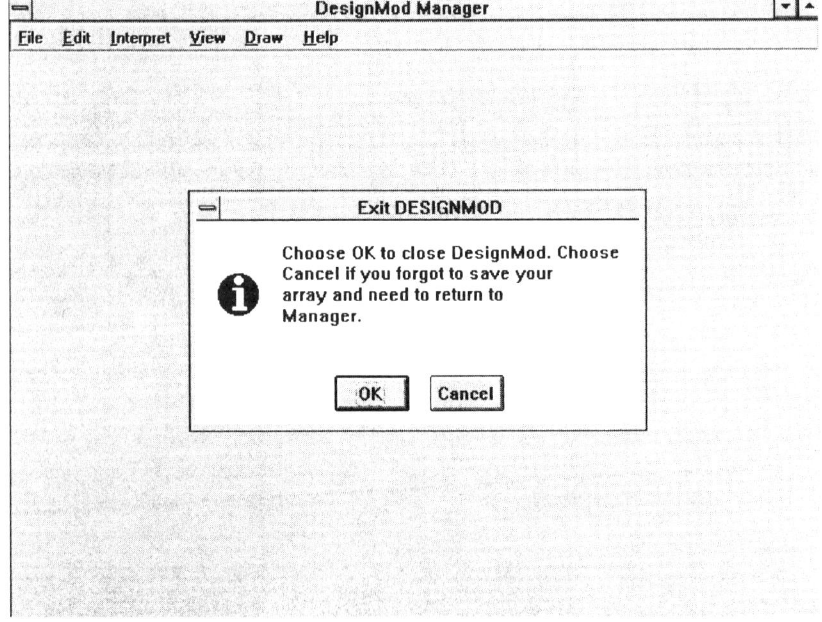

Figure B3.   Manager window at exit.

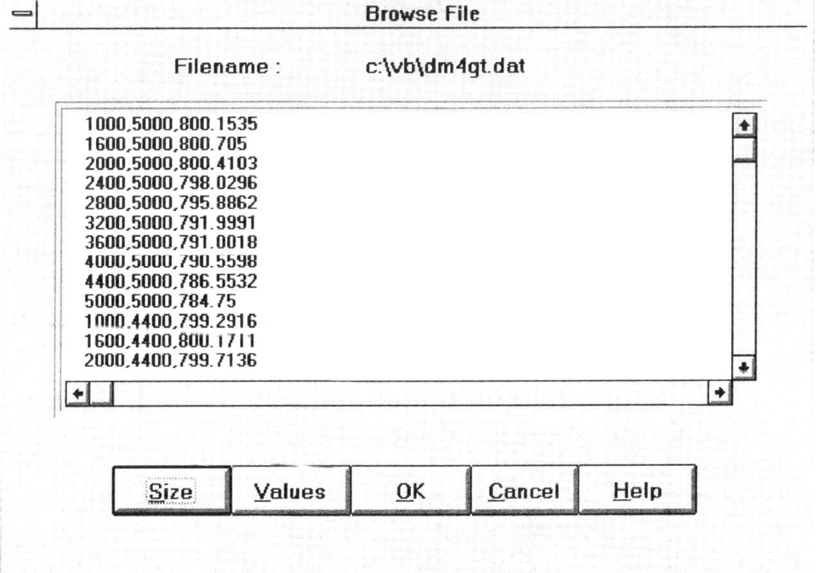

Figure B4.   File Catalog window with file names and descriptions.

Figure B5.   Browse File window with file contents.

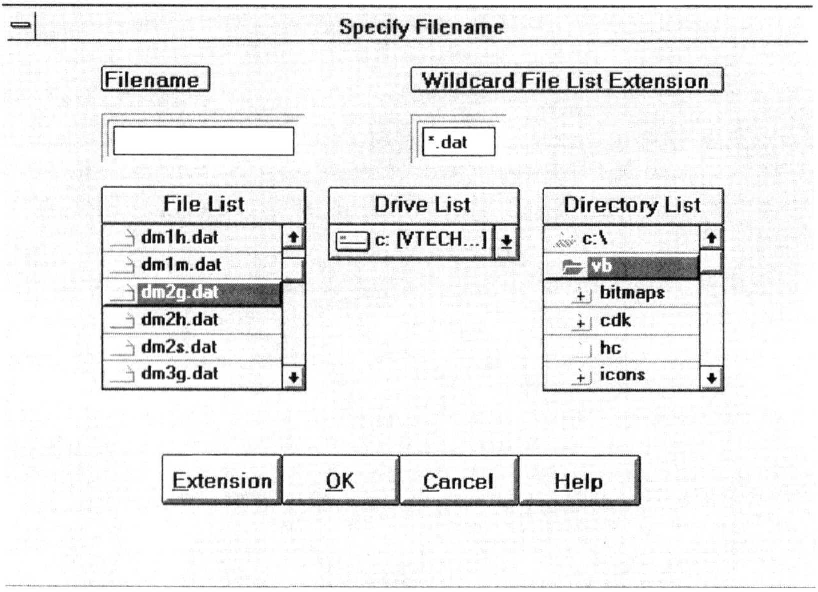

Figure B6.   Load File window with selected data filename.

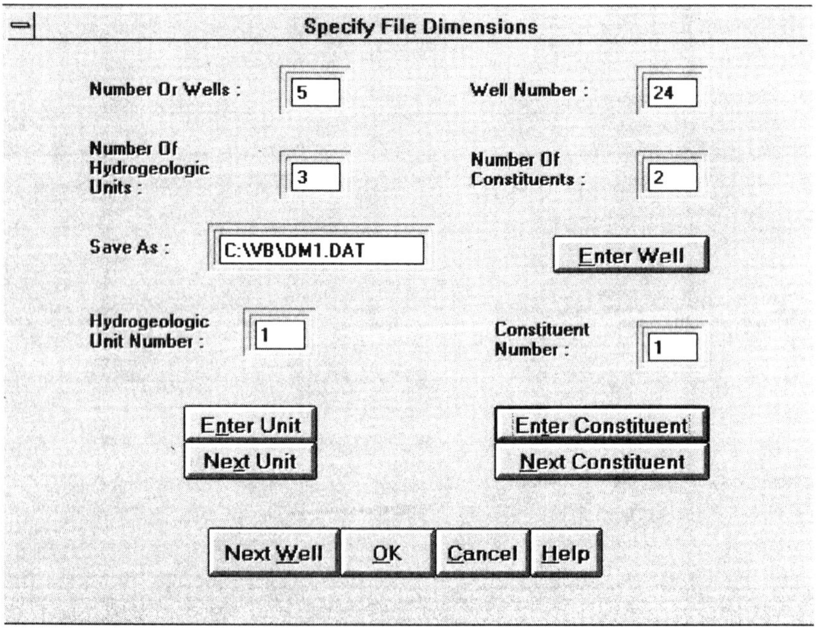

Figure B7.   Save Well database window with file dimensions.

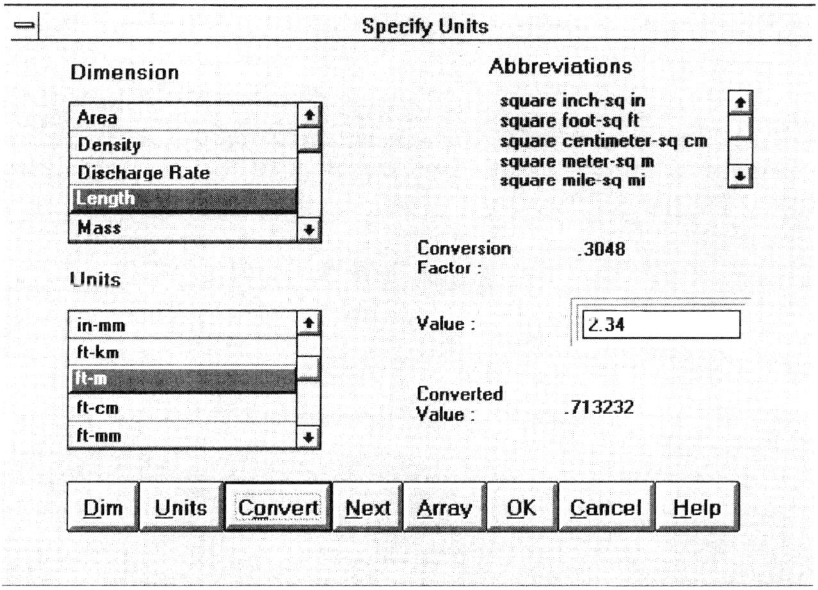

Figure B8.   Unit conversion window with selected dimension.

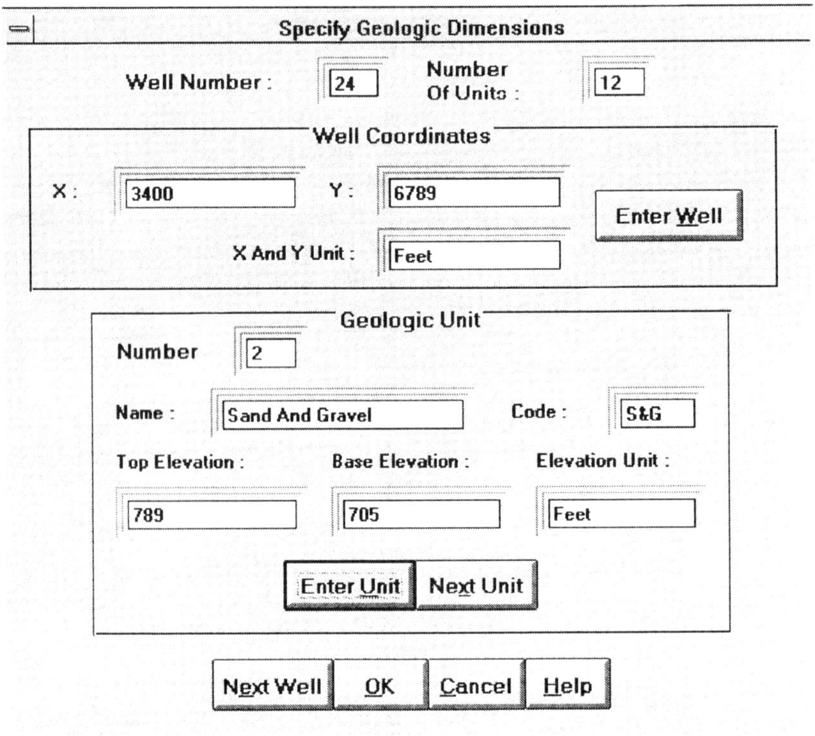

Figure B9.   Well geology window with database entries.

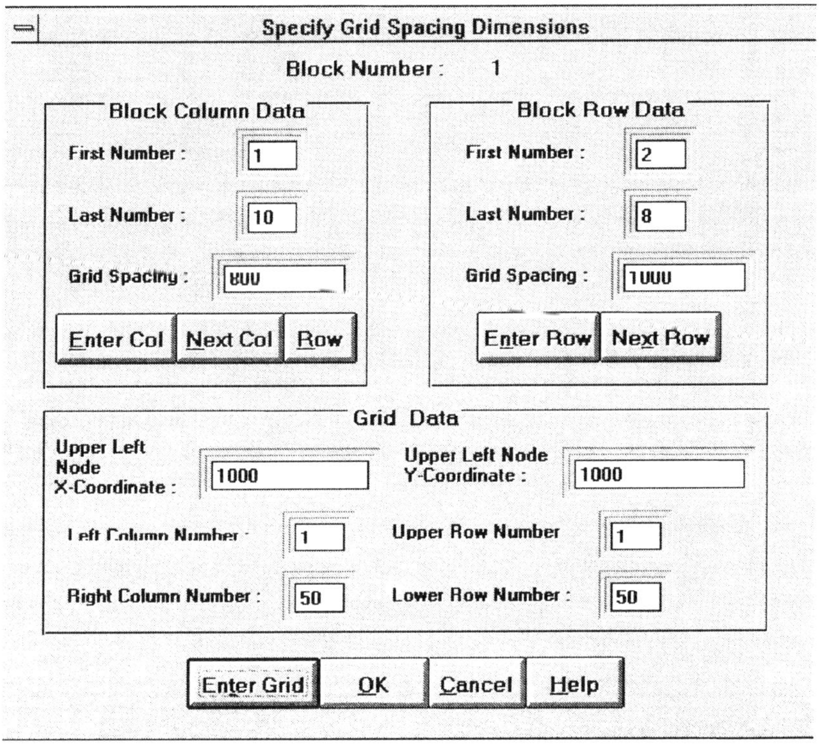

Figure B10.   Model grid window with database entries.

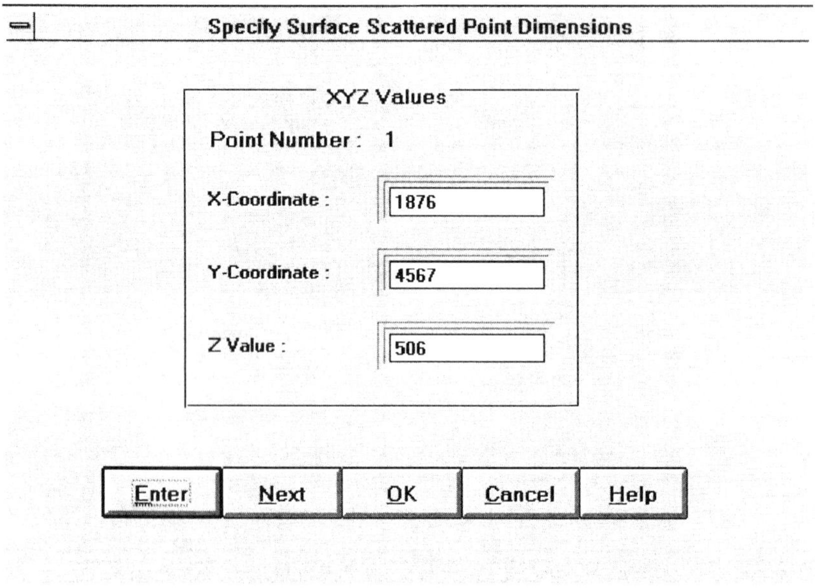

Figure B11.   Scattered surface point window with database entries.

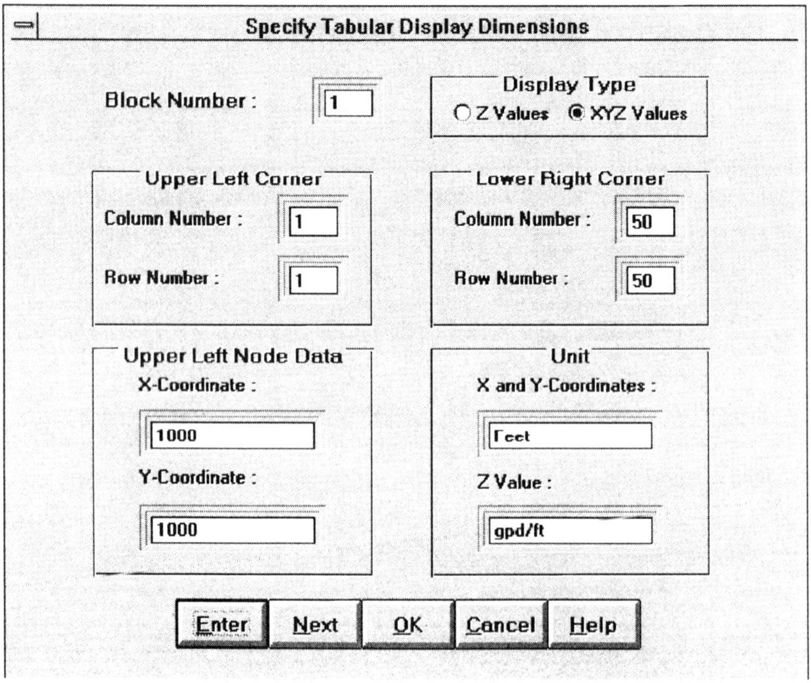

Figure B12.   Tabular Display window with dimension enties.

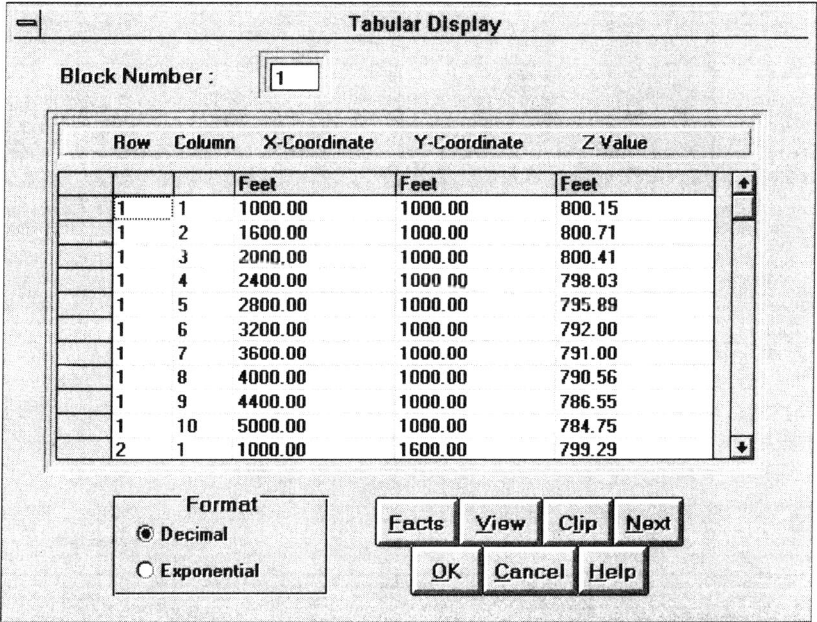

Figure B13.   Tabular Display window with spreadsheet contents.

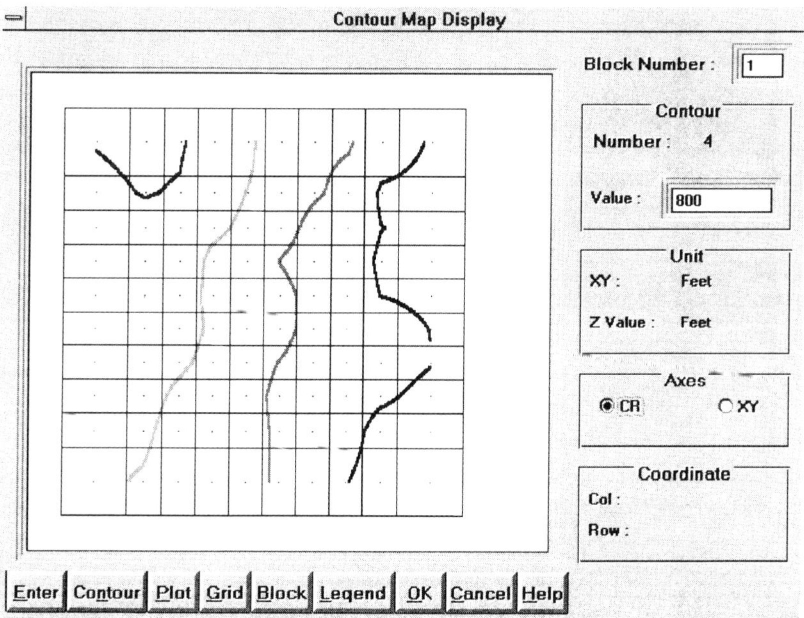

Figure B14.   Map Display window with contours.

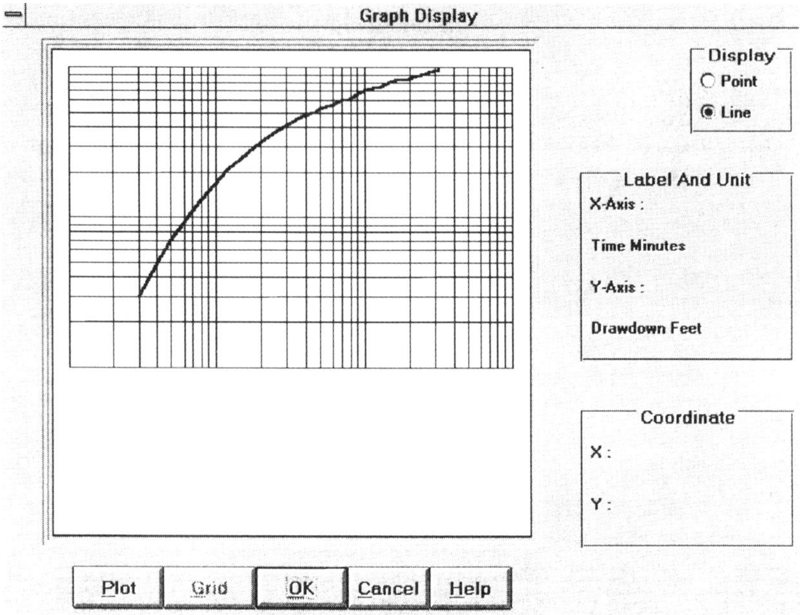

Figure B15. Graph Display window with logarithmic graph.

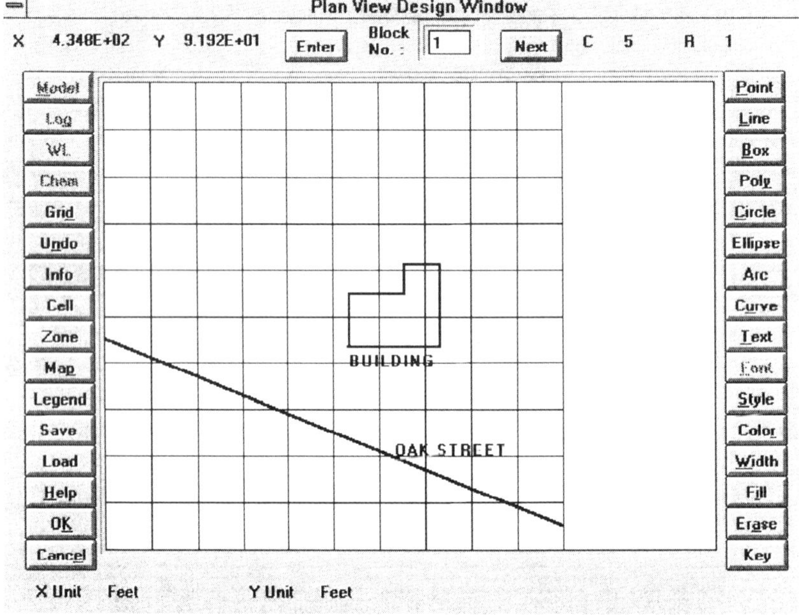

Figure B16. Design window with area of interest plan view features.

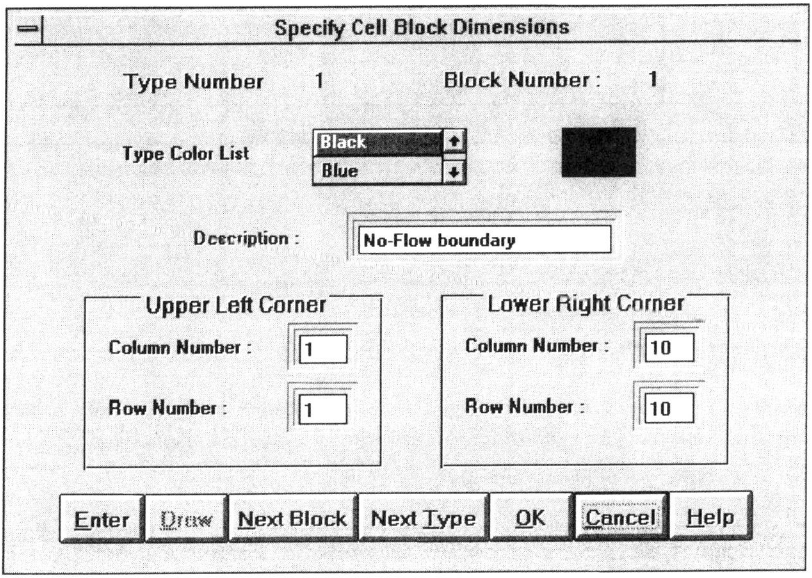

Specify Cell Block Dimensions

Type Number   1   Block Number :   1

Type Color List   Black
                         Blue

Description :   No-Flow boundary

Upper Left Corner
Column Number :   1

Row Number :   1

Lower Right Corner
Column Number :   10

Row Number :   10

Enter   Draw   Next Block   Next Type   OK   Cancel   Help

Figure B17.   Model grid cell color window with dimension entries.

Specify Box Dimensions

Coordinate Type
○ Screen   ⦿ Site

Upper Left Corner Site
X-Coordinate :   25

Upper Left Corner Site
X-Coordinate :   25

Lower Right Corner Site
Y-Coordinate :   500

Lower Right Corner Site
Y-Coordinate :   500

Draw   Next   OK   Cancel   Help

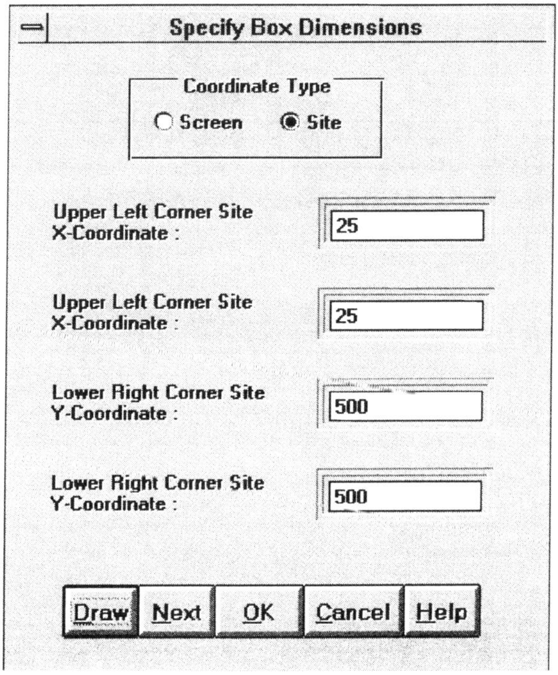

Figure B18.   Draw a box window with dimension entries.

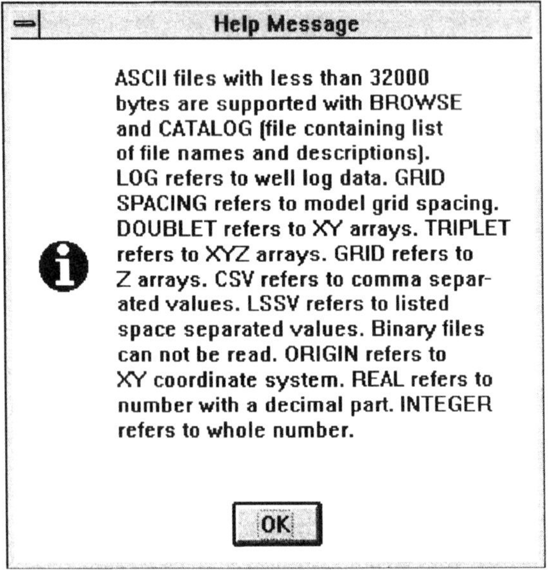

Figure B19.   Help window with typical message.

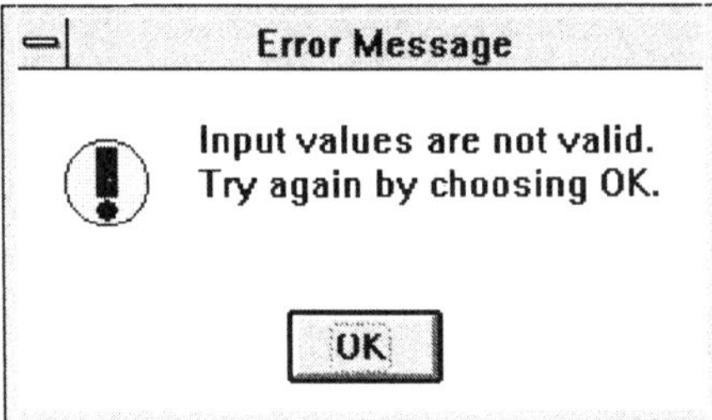

Figure B20.   Error window with typical message.

INDEX

A

Access for Windows, 17–18

aggregation in model databases, 10

algorithms for contours, 78–79

Analysis ToolPak, 17

anisotropy, 11

API (application programming
interface), 23–30

application programming interface (API)
functions of *Windows,* 23–30
with *Visual Basic,* 23–24

approximation in model databases, 10

aquifers, 109–112, 113, 124–126

Arc tool, 90

area, calculation of, within vertices,
153–154

Area on Interpret menu, 66, 68, 153–154

arithmetic graphs, 82–83, 147–148

arithmetic mean, 66, 68–69, 153–154

array. *See* data arrays

ASCII-formatted files, 5, 54

AUTOEXEC.BAT, editing in
Windows System Editor, 46

averaging in model databases, 10

axis options for graph design. *See also*
Graph on View menu
grid spacings, 83
labeling, 149
offset, 119

B

background literature on modeling, 7

backup diskette, 45

BASICA, 21

binary-formatted files, 5, 54
saving file as, 57, 97–98

blocks, 60–61

borders, 121–123

boundaries
constant head, 12, 112, 126–127
constant specified flux, 12
constant underflow, 112
flux, 9
impermeable, 9, 11
no-flow, 112, 125–126

Box tool, 87

Browse on File Menu, 53, 134–135, 177

building borders, 121–122

C

calculator, using *Windows* Calculator, 38

Cancel, to close dialog box, 42

Catalog on File menu, 51–52, 157–158
hard copies of, 52

cells
of finite-difference model grid, 60,
132, 156, 185
selecting range of, 42–43

Cell tool, 93, 94–95

chemical contamination, 111, 112, 113

Chem tool, for geochemical data
display, 92

Circle tool, 87–88, 89

C language-oriented programming
terminology, 23, 30–31

clicking the mouse, 39

clipboard
incorporation from *DesignMod* to
Word, 14
incorporation from *Windows* to
DesignMod, 43

Clipboard Viewer, using *Windows,* 43